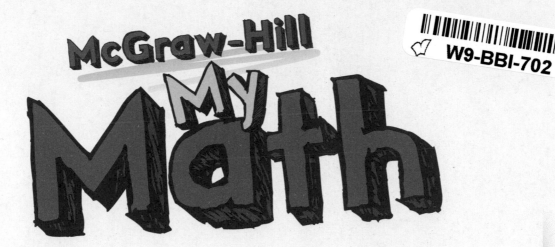

McGraw-Hill

My Math

Welcome to *My Math* – your very own math book!
You can write in it – in fact, you are encouraged to
write, draw, circle, explain, and color as you explore
the exciting world of mathematics. Let's get started.
Grab a pencil and finish each sentence.

My name is ___Anirudh.T___.

My favorite color is ___Red___.

My favorite hobby or sport is ___tennis___.

My favorite TV program or video game is
___Lego Marvel Super hero or Skylander___.

My favorite class is ___Math or Science___.

Mc
Graw
Hill
Education

mhmymath.com

STEM McGraw-Hill is committed to providing
instructional materials in Science, Technology, Engineering, and
Mathematics (STEM) that give all students a solid foundation,
one that prepares them for college and careers in the 21st
century.

Send all inquiries to:
McGraw-Hill Education
8787 Orion Place
Columbus, OH 43240

ISBN: 978-0-07-905765-5 (**Volume 1**)
MHID: 0-07-905765-9

Printed in the United States of America.

3 4 5 6 7 8 9 QSX 23 22 21 20 19 18

Understanding by Design® is a registered trademark of the Association for Supervision and
Curriculum Development ("ASCD").

McGraw-Hill My Math

Grade 5 • Volume 1

Authors:

Carter • Cuevas • Day • Malloy

Altieri • Balka • Gonsalves • Grace • Krulik • Molix-Bailey

Moseley • Mowry • Myren • Price • Reynosa • Santa Cruz

Silbey • Vielhaber

Mc
Graw
Hill
Education

GO digital ▶ connectED.mcgraw-hill.com

▶ Log In

1 Go to **connectED.mcgraw-hill.com.**

2 Log in using your username and password.

3 Click on the Student Edition icon to open the Student Center.

▶ Go to the Student Center

4 Click on Menu, then click on the **Resources** tab to see all of your online resources arranged by chapter and lesson.

5 Click on the **eToolkit** in the Lesson Resources section to open a library of eTools and virtual manipulatives.

6 Look here to find any assignments or messages from your teacher.

7 Click on the **eBook** to open your online Student Edition.

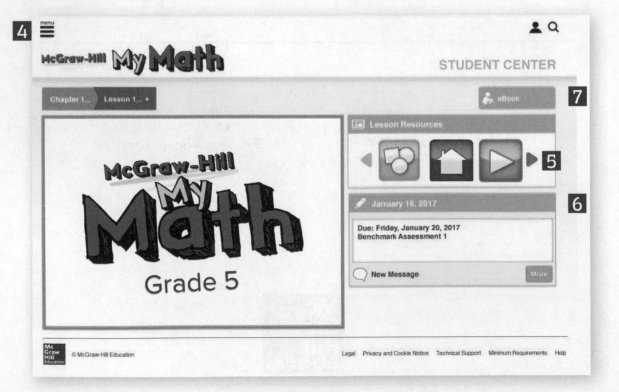

▶ Explore the eBook!

8 Click the **speaker icon** at the top of the eBook pages to hear the page read aloud to you.

More resources can be found by clicking the icons at the bottom of the eBook pages.

 Practice and review your Vocabulary.

 Animations and videos allow you to explore mathematical topics.

 Explore concepts with eTools and virtual manipulatives.

 Personal Tutors are online virtual teachers that walk you through the steps of the lesson problems.

 eHelp helps you complete your homework.

9

 Explore these fun digital activities to practice what you learned in the classroom.

 Worksheets are PDFs for Math at Home, Problem of the Day, and Fluency Practice.

Contents in Brief
Organized by Domain

Processes &Practices ➜ Woven Throughout

connectED.mcgraw-hill.com

Chapter

Place Value

ESSENTIAL QUESTION
How does the position
of a digit in a number relate
to its value?

Getting Started

Lessons and Homework

Wrap Up

Look for this!
There are
Brain Builder
problems in
every lesson.

connectED.mcgraw-hill.com

Number and Operations in Base Ten

ESSENTIAL QUESTION
What strategies can be used
to multiply whole numbers?

Chapter

2

Multiply Whole Numbers

Getting Started

Lessons and Homework

Wrap Up

Chapter

3 Divide by a One-Digit Divisor

Number and Operations in Base Ten

ESSENTIAL QUESTION
What strategies can be used
to divide whole numbers?

Getting Started

Lessons and Homework

Wrap Up

Look for this! eHelp
Click online and you
can get more help while
doing your homework.

Number and Operations in Base Ten

ESSENTIAL QUESTION
What strategies can I use to divide by a two-digit divisor?

Chapter

4 Divide by a Two-Digit Divisor

Getting Started

Lessons and Homework

Wrap Up

Look for this! Tools
Click online and you can find tools that will help you explore concepts.

x

connectED.mcgraw-hill.com

Number and Operations in Base Ten

ESSENTIAL QUESTION
How can I use place value and properties to add and subtract decimals?

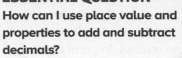

Chapter

5 Add and Subtract Decimals

Getting Started

Lessons and Homework

Wrap Up

connectED.mcgraw-hill.com

Number and Operations in Base Ten

ESSENTIAL QUESTION
How is multiplying and dividing decimals similar to multiplying and dividing whole numbers?

Chapter

6

Multiply and Divide Decimals

Getting Started

Lessons and Homework

Wrap Up

connectED.mcgraw-hill.com

Operations and Algebraic Thinking

ESSENTIAL QUESTION
How are patterns used
to solve problems?

Chapter

7 Expressions and Patterns

Getting Started

Lessons and Homework

Wrap Up

Look for this! **Tutor**
Click online and you
can watch a teacher
solving problems.

ESSENTIAL QUESTION
How are factors and multiples helpful in solving problems?

Chapter

8 Fractions and Decimals

Getting Started

Lessons and Homework

Wrap Up

Look for this!

Vocab abc

Click online and you can find activities to help build your vocabulary.

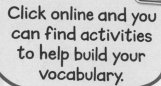

connectED.mcgraw-hill.com

Chapter

9 Add and Subtract Fractions

Getting Started

Lessons and Homework

Wrap Up

Number and Operations — Fractions

ESSENTIAL QUESTION
What strategies can be used to multiply and divide fractions?

Chapter

10 Multiply and Divide Fractions

Getting Started

Lessons and Homework

Wrap Up

connectED.mcgraw-hill.com

Chapter 11 Measurement

Measurement and Data

ESSENTIAL QUESTION
How can I use measurement conversions to solve real-world problems?

Getting Started

Lessons and Homework

Wrap Up

connectED.mcgraw-hill.com

Geometry

ESSENTIAL QUESTION
How does geometry help
me solve problems in
everyday life?

Chapter

12 Geometry

Getting Started

Lessons and Homework

Wrap Up

connectED.mcgraw-hill.com

ESSENTIAL QUESTION

How does the position of a digit in a number relate to its value?

Let's Go Outdoors!

Watch a video!

Watch

MY Chapter Project

Map It!

Day 1

1. Work together to draw an outline map of the United States on poster board.

2. Use the Internet or the library to research the land area in square miles of each state and record that information on the map.

3. Think of a comparison question you could ask your classmates. *Sample questions: Which state has the greater number of square miles, Wisconsin or Iowa? Is the number of square miles of Louisiana and Arkansas together greater than or less than the number of square miles of Texas?*

Question: _____

_____

_____

Day 2

1. Work together to create a guide to your map by listing all the states in order of land area, from least to greatest, in a 2-column table or spreadsheet.

2. Ask the class your comparison question and answer other groups' comparison questions.

3. List the 10 largest states and their land area in square miles in the table below.

State	Land Area (square miles)

Name _____

Am I Ready?

Write each number in word form.

1. 8 _____

2. 23 _____

3. 15 _____

4. 160 _____

Write the number that represents each point on the number line.

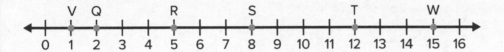

5. Q _____

6. S _____

7. R _____

8. T _____

9. V _____

10. W _____

Write each sentence using the symbols <, >, or =.

11. 8 is less than 12.

12. 24 is greater than 10.

_____   _____

13. The high temperature yesterday was 64°F.
The high temperature today is 70°F. Write
64 is less than 70 using the symbols <, >, or =.

_____

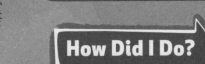

Shade the boxes to show the problems you answered correctly.

| 1 | 2 | 3 | 4 | 5 | 6 | 7 | 8 | 9 | 10 | 11 | 12 | 13 |

Online Content at connectED.mcgraw-hill.com

Name
....

MY Math Words
Vocab

Review Vocabulary

comma	hundreds	hundred thousands	ones
tens	ten thousands	thousands	

Making Connections
Use the review words to complete each section of the bubble map.

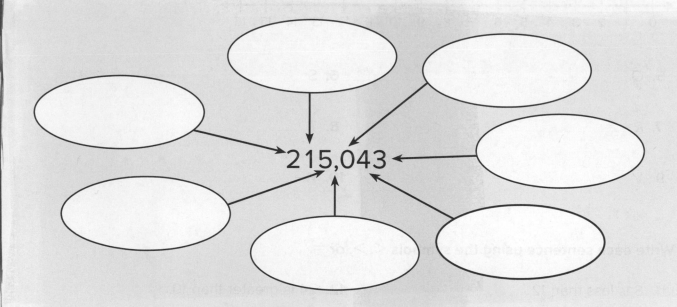

Describe how commas are used in writing greater numbers.

_____

_____

_____

Processes & Practices

Lesson 1–3

decimal point

1.378   5.0   6.78

Lesson 1–3

decimal

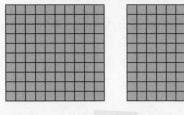

1.75

Lesson 1–7

equivalent decimals

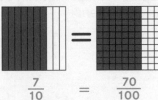

$$\frac{7}{10} = \frac{70}{100}$$

0.7 = 0.70

Lesson 1–1

expanded form

12,002,060 = 1 × 10,000,000 + 2 × 1,000,000 + 2 × 1,000 + 6 × 10

Lesson 1–1

period

Millions Period			Thousands Period			Ones Period		
Millions			Thousands			Ones		
hundreds	tens	ones	hundreds	tens	ones	hundreds	tens	ones
6	5	0	0	8	4	9	7	0

Lesson 1–1

place

Millions Period			Thousands Period			Ones Period		
Millions			Thousands			Ones		
hundreds	tens	ones	hundreds	tens	ones	hundreds	tens	ones
6	5	0	0	8	4	9	7	0

Lesson 1–1

standard form

3,000 + 400 + 90 + 1 = 3,491

standard form

Lesson 1–1

place value

Millions Period			Thousands Period			Ones Period		
Millions			Thousands			Ones		
hundreds	tens	ones	hundreds	tens	ones	hundreds	tens	ones
	5	0	0	8	0	0	0	0

5 = ten millions
8 = ten thousands

Ideas for Use

- Design a crossword puzzle. Use the definition for each word as the clues.

- Work with a partner to name the part of speech of each word. Consult a dictionary to check your answers.

A number that has a digit in the tenths place, hundredths place, and/or beyond.

Erin ate $\frac{3}{4}$ of a sandwich. Write the amount eaten as a decimal.

_____

A period separating the ones and the tenths in a decimal number.

How does a number with a decimal point differ from a number without one?

_____

_____

A way of writing a number as the sum of the values of its digits.

Expand can mean "to stretch or unfold." How can this definition help you remember this definition?

_____

_____

Decimals that have the same value.

Name two topics in math that can be described as *equivalent*.

_____

_____

The position of a digit in a number.

What is a synonym for *place*?

_____

The name given to each group of three digits in a place-value chart.

What is the meaning of *period* in language arts?

_____

_____

The value given to a digit by its place in a number.

Write a six-digit number. Then write the place value for each digit.

_____

_____

The usual way of writing a number that shows only its *digits,* no words.

Write a different meaning of the word *standard*.

_____

_____

MY Vocabulary Cards

Lesson 1–1

place-value chart

Millions Period			Thousands Period			Ones Period		
Millions			Thousands			Ones		
hundreds	tens	ones	hundreds	tens	ones	hundreds	tens	ones
6	5	0	0	8	4	9	7	0

Ideas for Use

- Use a blank card to write this chapter's essential question. Use the back of the card to write or draw examples that help you answer the question.

- Use blank cards to review key concepts from the chapter. Write a few study tips on the back of each card.

A chart showing the value of each number in a multi-digit whole number

Use the space below to write a place-value chart for 7,932,035.

MY Foldable

FOLDABLES® Follow the steps on the back to make your Foldable.

9 0 0 , 0 0 0 , 0 0 0

hundred
millions

8 0 0 , 0 0 0 , 0 0 0

ten
millions

7 , 0 0 0 0 , 0 0 0

millions

6 _ _ _ , _ _ _ _

hundred
thousands

5 _ _ , _ _ _ _

ten
thousands

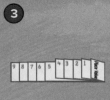

Place Value

1 ones

2 ___ tens

3 ___ ___ hundreds

4 , ___ ___ ___ thousands

Lesson 1
Place Value Through Millions

ESSENTIAL QUESTION
How does the position of a digit in a number relate to its value?

A **place-value chart** shows the value of the digits in a number. In greater numbers, each group of three digits is separated by commas, and is called a **period**.

 Math in My World Watch Tools Tutor

Example 1

The distance from Earth to the Sun is **92,955,793 miles.** Use the place-value chart to list the value of each digit.

1 Complete the place-value chart.

Millions Period			Thousands Period			Ones Period		
hundreds	tens	ones	hundreds	tens	ones	hundreds	tens	ones

2 List the values of each digit.

$9 \times 10,000,000 \rightarrow 90,000,000$

$2 \times 1,000,000 \rightarrow 2,000,000$

$9 \times 100,000 \rightarrow$ _____

$5 \times 10,000 \rightarrow$ _____

$5 \times 1,000 \rightarrow$ _____

$7 \times 100 \rightarrow$ _____

$9 \times 10 \rightarrow$ _____

$3 \times 1 \rightarrow$ _____

The digit 5 in the ten thousands place is 10 times greater than the digit 5 in the thousands place.

A digit in one **place**, or **place value**, represents 10 times as much as it represents in the place to its right and $\frac{1}{10}$ of what it represents in the place to its left.

The **standard form** of a number is the usual or common way to write a number using digits. The **expanded form** of a number is a way of writing a number as the sum of the values of its digits. The places with zero as a digit are not included in the expanded form.

Example 2

Tutor

The human eye blinks an average of 5,500,000 times a year. Write 5,500,000 in word form and expanded form.

 Write the number in the place-value chart.

Millions Period			Thousands Period			Ones Period		
hundreds	tens	ones	hundreds	tens	ones	hundreds	tens	ones

Write the number in word form.

five _____ , *five* _____ *thousand*

Write the number in expanded form.
five million: 5 × 1,000,000
five hundred thousand: 5 × 100,000

In expanded form, 5,500,000 =

_____ × _____ + _____ × _____

Guided Practice

Write the value of the highlighted digit.

1. 469,999 _____

2. 35,098,098 _____

3. Circle the digit in the ten thousands place.

    1, 2 3 5, 9 8 0

Talk MATH

Explain how the value of the highlighted digit in the number 26,077,928 compares to the digit to its left.

Independent Practice

Write the value of the highlighted digit.

4. 3,132,685 _____

5. 5,309,573 _____

6. 1,309,841 _____

Write each number in word form and expanded form.

7. 5,901,452 _____

_____

_____

8. 309,099,990 _____

_____

Write each number in standard form and expanded form.

9. *eighty-three million, twenty-three thousand, seven*

_____

_____

10. *three hundred four million, eight hundred thousand, four hundred*

_____

_____

Use the place-value chart for Exercises 11 and 12.

Millions Period			Thousands Period			Ones Period		
hundreds	tens	ones	hundreds	tens	ones	hundreds	tens	ones
		5	9	0	1	4	5	2

11. The 9 is in the _____ place.

12. The 1 has a value of 1 × _____ .

Problem Solving

13. In a recent year, the population of the United States was about 304,967,000. Write the population in word form.

_____

_____

_____

14. The land area of Florida is 1 × 100,000 + 3 × 10,000 + 9 × 1,000 + 8 × 100 + 5 × 10 + 2 × 1 square kilometers. Write the area in standard form and word form.

_____

_____

_____

 ## Brain Builders

15. **Processes &Practices**  **Explain to a Friend**

The amount of time that American astronauts have spent in space is about 13,507,804 minutes. What is the word form of that number?

_____

_____

_____

16. **Processes &Practices** **Use Number Sense**

Write the number with the least value using the digits 1 through 9. Use each digit only once. Explain your reasoning.

_____

_____

_____

_____

17. **Building on the Essential Question** Explain how you know what number is missing in the equation 3,947 = 3,000 + ■ + 40 + 7.

_____

_____

_____

_____

MY Homework

Homework Helper

Need help? connectED.mcgraw-hill.com

Use the place-value chart to write 12,498,750 in word form and expanded form.

1 Write the number in the place-value chart.

Millions Period			Thousands Period			Ones Period		
hundreds	tens	ones	hundreds	tens	ones	hundreds	tens	ones
	1	2	4	9	8	7	5	0

2 Write the number in word form.

twelve million, four hundred ninety-eight thousand, seven hundred fifty

3 Write the number in expanded form.

1 × 10,000,000 + 2 × 1,000,000 + 4 × 100,000 +
9 × 10,000 + 8 × 1,000 + 7 × 100 + 5 × 10

Practice

Write the value of the highlighted digit.

1. 1,283,479

2. 50,907,652

3. 318,472,008

4. Write 103,727,495 in word form and expanded form.

5. Hanna stated that 11,760,825 people saw the Miami Heat play
last season. Chris wants to be sure he heard the number correctly.
Write 11,760,825 in word form and expanded form for Chris.

_____

_____

_____

Brain Builders

6. **Processes &Practices** **Find the Error** American car makers produce
5,650,000 cars each year. In a report, Ben wrote that Americans
made 6,550,000 cars. What mistake did Ben make? How can he fix it?

_____

_____

_____

Vocabulary Check [Vocab abc]

Match the vocabulary word with its definition.

7. standard form

• each group of three digits on
a place-value chart

8. period

• a way of writing a number as the sum of the
values of its digits

9. expanded form

• the usual or common way to write
a number using digits

10. **Test Practice** A popular pirate movie made $135,634,554 in sales during one
weekend. What is $\frac{1}{10}$ the value of the highlighted digit?

Ⓐ $30,000

Ⓒ $300,000

Ⓑ $3,000,000

Ⓓ $30,000,000

Name

....

ESSENTIAL QUESTION
How does the position of a digit in a number relate to its value?

To compare numbers, you can use place value and the symbols <, >, and =.

Words	Symbol
is greater than	>
is less than	<
is equal to	=

Math in My World Watch Tools Tutor

Example 1

The table shows the two largest oceans in the world. Which ocean has a greater area?

Ocean	Approximate Area (square miles)
Atlantic Ocean	33,420,160
Pacific Ocean	64,186,600

1 Write the numbers in the place-value chart.

Millions Period			Thousands Period			Ones Period			
hundreds	tens	ones	hundreds	tens	ones	hundreds	tens	ones	
									← Atlantic Ocean
									← Pacific Ocean

2 Begin at the greatest place. Compare the digits.

6 ◯ 3

Since 6 is _____ than 3, then 64,186,600 > 33,420,160.

So, the _____ has a greater area.

Copyright © McGraw-Hill Education ©singularone/iStock/Getty Images

Example 2 Tutor

The table shows the area in square miles in different countries. Use place value to order the countries from *greatest* area to *least* area.

Country Areas	
Country	Area (square miles)
Argentina	1,068,296
Australia	2,967,893
India	1,269,338
Norway	125,181

1 Line up the ones place. Compare the digits in the greatest place.

2 > _____

So, _____ has the greatest area.

1,068,296
2,967,893
1,269,338
125,181

2 Compare the digits in the next place.

2 > 0.

So, _____ has the next greatest area.

3 Since 1,068,296 > _____,

the area of _____ is greater

than the area of _____.

The order from *greatest* area to *least* area is Australia, _____,

_____, and _____.

Guided Practice

Write <, >, or = in each ◯ to make a true sentence.

1. 655,543 ◯ 556,543

2. 10,027,301 ◯ 10,207,301

3. Order the numbers 145,099; 154,032; 145,004; and 159,023 from *greatest* to *least*.

_____

Talk MATH

When ordering whole numbers, explain what to do when the digits in the same place have the same value.

Name ....

Independent Practice

Write $<$, $>$, or $=$ in each ◯ to make a true sentence.

4. 462,211 ◯ 426,222

5. 42,235,909 ◯ 42,324,909

6. 20,318,523 ◯ 21,318,724

7. 96,042,317 ◯ 96,042,317

8. 132,721,424 ◯ 132,721

9. 152,388,000 ◯ 152,388,010

10. 113,222,523 ◯ 113,333,523

11. 767,676,767 ◯ 676,767,676

Order the numbers from _greatest_ to _least_.

12. 138,023; 138,032; 139,006; 183,487

_____

13. 3,452,034; 4,935,002; 34,035,952; 34,530,953

_____

14. 731,364,898; 731,643,898; 73,264,898; 731,643,989

_____

15. 395,024,814; 593,801,021; 395,021,814; 39,021,814

_____

Order the numbers from _least_ to _greatest_.

16. 85,289,688; 85,290,700; 85,285,671; 85,301,001

_____

17. 32,356,800; 33,353,800; 32,937,458; 33,489,251

_____

18. 2,009,146; 2,037,579; 2,006,981; 2,011,840

_____

19. 854,236,100; 855,963,250; 855,903,675; 854,114,370

_____

Copyright © McGraw-Hill Education

Problem Solving

20. Rank the following states from *least* to *greatest* population.

State Population	
State	**Population**
Alabama	4,627,851
Colorado	4,861,515
Mississippi	2,918,785
Ohio	11,466,917

_____

_____

_____

21. Order the cars from most expensive to least expensive.

Most Expensive Cars	
Car	**Price ($)**
Bugatti Veyron 16.4	1,192,057
Leblanc Mirabeau	645,084
Pagani Zonda Roadster	667,321
Saleen S7	555,000

_____

_____

_____

Brain Builders

22. **Processes &Practices** **2** **Reason** Write five numbers that are greater than 75,330,000 but less than 75,410,000.

_____

_____

23. **Building on the Essential Question** How do you compare whole numbers through the millions?

_____

_____

_____

Name

Homework Helper

Need help? connectED.mcgraw-hill.com

Order the following numbers from *least* to *greatest*.

84,189,688; 85,290,700; 58,285,671; 80,301,785

1 Line up the ones place. Compare the digits in the greatest place.

5 < 8
So, 58,285,671 is the least.

84,189,688
85,290,700
58,285,671
80,301,785

2 Compare the digits in the next place.
80,301,785 < 84,189,688

3 85,290,700 > 84,189,688
So, 85,290,700 is the greatest number.

The order from least to greatest is 58,285,671; 80,301,785; 84,189,688; and 85,290,700.

Practice

Write <, >, or = in each ◯ to make a true sentence.

1. 67,982,001 ◯ 67,892,001

2. 100,542,089 ◯ 105,042,098

3. 1,986,034 ◯ 1,896,075

4. 12,165,982 ◯ 12,178,983

5. 239,742,005 ◯ 289,650,010

6. 1,652,985 ◯ 1,563,218

Order the numbers from *greatest* to *least*.

7. 3,356,000; 2,359,412; 2,937,158; 3,368,742

8. 2,009,832; 2,103,425; 2,009,604; 2,112,300

Order the numbers from *least* to *greatest*.

9. 14,258,123; 14,259,688; 14,256,001; 14,258,252

_____

10. 574,210,033; 574,211,874; 574,198,852; 874,210,089

_____

Problem Solving

11. The list below shows how many kids play each sport. Order the sports from most players to least players.

   Soccer: 3,875,026   Surfing: 250,982

   Baseball: 900,765   Basketball: 2,025,351

_____

12. Andrea wants to live in the city with the most people. She read that New York City has 8,008,278 people and that Seoul, South Korea has 10,231,217 people. In which city does Andrea want to live? Explain how you know.

_____

13. The Denver Mint made 2,638,800 nickels. The Philadelphia Mint made 2,806,000 nickels. Which mint made more nickels? Explain how you know.

_____

Brain Builders

14. **Processes &Practices** **3** **Draw a Conclusion** In 1950, bike stores sold about 205,850 bikes per year. In 2000, bike stores sold about 185,000 bikes per year. Is the number of bikes being sold getting larger or smaller? Explain how you know.

_____

15. **Test Practice** Which set of numbers are in order from *greatest* to *least*?

   Ⓐ 74,859,623; 74,759,458; 74,905,140; 73,569,991

   Ⓑ 74,905,140; 74,859,623; 74,759,458; 73,569,991

   Ⓒ 73,569,991; 74,759,458; 74,859,623; 74,905,140

   Ⓓ 74,905,140; 74,759,458; 74,859,623; 73,569,991

Name _____

Numbers that have digits in the tenths place, hundredths place, and/or beyond are called **decimals**. A **decimal point** is used to separate the ones place from the tenths place.

Draw It

Use a model to show $\frac{3}{10}$. Then write it in word form and as a decimal.

1 Shade _3/10_ columns of the tenths model.

2 The model shows:

Word form: _three_ tenths

Decimal: 0. [3]

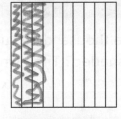

Ones	Tenths
0	3

Try It

Use a model to show $\frac{9}{100}$. Then write it in word form and as a decimal.

1 Shade _9/100_ of the small squares on the hundredths model.

2 The model shows:

Word form: _nine_ hundredths

Decimal: 0. [0] [9]

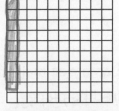

Ones	Tenths	Hundredths
0	0	9

Try It

Use a model to show $\frac{34}{100}$. Then write it in word form and as a decimal.

1 Shade __34/DO__ of the small squares.

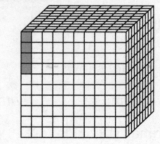

2 The model shows:

Word form: _thirty four_ hundredths

Notice that there are __3__ tenths

and __4__ hundredths shaded.

Decimal: 0. [3] [4]

Ones	Tenths	Hundredths
0	3	4

Talk About It

1. The model at the right shows a thousandths cube. What fraction of the model is shaded? Then write it as a decimal.

_0.004_

2. **Processes &Practices** **5** **Use Math Tools** Shade the model to show $\frac{80}{100}$. Then write the fraction in word form and as a decimal.

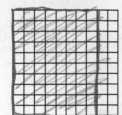

Zero and eight tenths

3. Explain why $\frac{45}{100}$ is written as a decimal with a 4 in the tenths place and a 5 in the hundredths place.

It is because, if 45 is split it would be 40 and 5, 5 = five hundedth and 40 = 4 thenths.

Practice It

Shade the model to show each fraction. Write each fraction in word form.

4. $\frac{7}{10}$

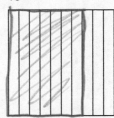

Seven thenths

5. $\frac{9}{10}$

nine tenths

6. $\frac{4}{10}$

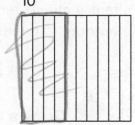

four thenths

7. $\frac{5}{100}$

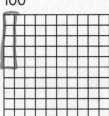

five hundredth

8. $\frac{63}{100}$

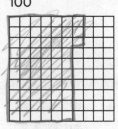

sixty three hundredth

9. $\frac{21}{100}$

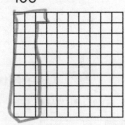

twenty one hundredths

Write the decimal for each model.

10.

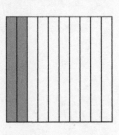

0.2

11.

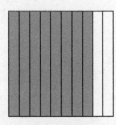

0.8

12.

0.5

13.

0.17

14.

6.9

15.

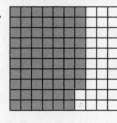

0.68

Apply It

16. Dewayne waters his plant with 0.76 quart of water every other day. Write the amount of water he uses in word form and as a fraction.

Seventy six hundredths $\frac{76}{100}$

17. Jessica's family ate six-tenths of the cake she baked. Write the amount eaten as a fraction and a decimal.

0.6 $\frac{6}{10}$

18. **Processes &Practices** **3** **Draw a Conclusion** Mei rode her bike $\frac{7}{10}$ mile to school. Her friend Julie walked 0.07 mile to school. Did the two girls travel the same distance? Explain.

NO because $\frac{7}{10}$ as a decimal is 0.7 and 0.07 < 0.7 so it is not =.

19. **Processes &Practices** **2** **Use Number Sense** Write the decimal and fraction for each model. Are the two models equal? Explain.

$\frac{4}{10}$ 0.4   0.40

yes because the two digits are the same then you can add as many 0 as you want so he added one.

Write About It

20. How do you model a decimal using a fraction?

You can model it by using the place value like 0.47 it has two digit so $\frac{47}{100}$.

26 **Chapter 1** Place Value

Name

MY Homework

Homework Helper

Need help? connectED.mcgraw-hill.com

Use a model to show $\frac{57}{100}$. Then write it in word form and as a decimal.

1. Shade 57 of the small squares.

2. The model shows:

   Word form: *fifty-seven hundredths*
   Decimal: 0.57

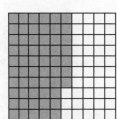

Practice

Shade the model to show each fraction. Write each fraction in word form.

1. $\frac{4}{10}$

four tenths

2. $\frac{8}{10}$

eight tenths

3. $\frac{45}{100}$

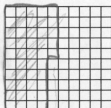

fourty five hundredths

4. $\frac{76}{100}$

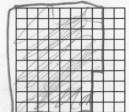

Seventy six hundredths

Write the decimal for each model.

5.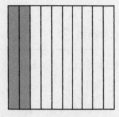

__0.2__

6.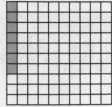

__0.07__

7.

__0.9__

8.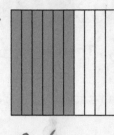

__0.6__

 # Problem Solving

9. **Processes &Practices** **Use Number Sense** The largest butterfly in the world is found in Papua, New Guinea. The female of the species weighs about $\frac{9}{10}$ ounce. Use a decimal to write the female's weight.

__0.9 or nine tenths oz.__

Vocabulary Check

10. Choose the correct word(s) to complete the sentence below.
    decimal   decimal point

    A _____ is used to separate the ones place from the tenths place.

11. Circle all of the following that represent $\frac{32}{100}$.

    thirty-two hundredths

    0.032

    0.32

    thirty-two tenths

Lesson 4
Represent Decimals

ESSENTIAL QUESTION
How does the position of a digit in a number relate to its value?

 Math in My World | Watch ▶ | Tutor 💬

Example 1

A bee hummingbird weighs only about $\frac{56}{1,000}$ of an ounce. Represent this fraction as a decimal. Then write it in word form.

1 The model represents thousandths by showing one thousand small cubes.

Shade ___56___ of the small cubes.

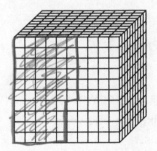

HUMMING ALONG!

2 The fraction names thousandths, so there should be three digits to the right of the decimal point.

$\frac{56}{1,000} = 0.$ | 0 | 5 | 6 |

3 Write $\frac{56}{1,000}$ in word form. $\frac{56}{1,000}$ ← fifty-six
                                                         ← thousandths

So, $\frac{56}{1,000}$ is ___Fifty six thousandth___

Example 2

Model $\frac{35}{100}$. **Then write it in word form and as a decimal.**

1 Shade _____ of the small squares.

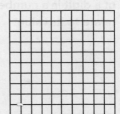

2 The fraction names *thirty-five*

_____, so there should be two digits to the right of the decimal point.

So, $\frac{35}{100}$ is *thirty-five hundredths*

and 0. ☐ ☐ .

Guided Practice

Shade the model. Then write each fraction as a decimal.

1. $\frac{2}{10}$

Decimal: _____

2. $\frac{58}{100}$

Decimal: _____

Describe a rule for writing fractions like $\frac{8}{100}$ and $\frac{32}{1,000}$ as decimals.

3. $\frac{95}{1,000}$

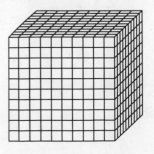

Decimal: _____

Independent Practice

Shade the model. Then write each fraction in word form and as a decimal.

4. $\frac{3}{10}$

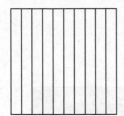

Word form: _____

Decimal: _____

5. $\frac{86}{100}$

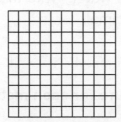

Word form: _____

Decimal: _____

6. $\frac{99}{100}$

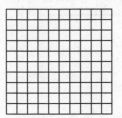

Word form: _____

Decimal: _____

7. $\frac{51}{1,000}$

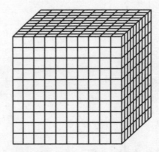

Word form: _____

Decimal: _____

8. $\frac{22}{1,000}$

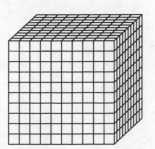

Word form: _____

Decimal: _____

9. $\frac{1}{1,000}$

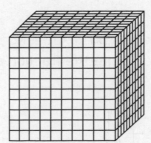

Word form: _____

Decimal: _____

Problem Solving

10. A runner decreased his time by $\frac{5}{100}$ second. Express the decrease as a decimal.

   _0.05_

11. **Processes &Practices** **2** **Use Number Sense** About $\frac{7}{10}$ of a person's body weight is water. Write this fraction in word form and as a decimal.

   _0.7_   _seven tenths_

Write the customary measure for each metric measure as a decimal.

12. 1 kilometer = _0.62 miles_

13. 1 millimeter = _0.04 in._

14. 1 gram = _0.035 oz._

15. 1 liter = _0.908 quart_

Metric Measure	Customary Measure
1 kilometer	$\frac{62}{100}$ mile
1 millimeter	$\frac{4}{100}$ inch
1 gram	$\frac{35}{1,000}$ ounce
1 liter	$\frac{908}{1,000}$ quart

Brain Builders

16. **Processes &Practices** **3** **Find the Error** Dylan is writing $\frac{95}{1,000}$ as a decimal. Find his mistake and correct it.

   $$\frac{95}{1,000} = 0.950$$

   The 95 in the decimal is suppose two be in the 5 and 0's place.

17. **Building on the Essential Question** How can the word form of a fraction help you write the fraction as a decimal?

   It can help you because fraction's word form is the same as decimal.

32 **Chapter 1** Place Value

Name ....

MY Homework

Homework Helper Need help? connectED.mcgraw-hill.com

Model $\frac{98}{1,000}$. Then write it in word form and as a decimal.

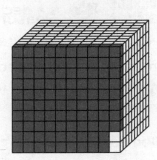

1 The model shows 98 of the small cubes are shaded.

2 The fraction names *ninety-eight thousandths*, so there should be three digits to the right of the decimal point.

So, $\frac{98}{1,000} = 0.098$.

Practice

Shade the model. Then write each fraction in word form and as a decimal.

1. $\frac{7}{10}$

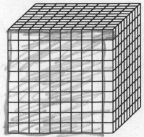

seven tenths

2. $\frac{62}{100}$

Sixty two hundredths

3. $\frac{91}{1,000}$

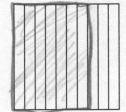

ninty one thousandths

4. $\frac{75}{1,000}$

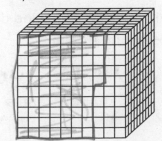

Seventy five thousandths

Write each fraction as a decimal.

5. $\frac{15}{100} =$ _0.15_

6. $\frac{129}{1,000} =$ _0.129_

7. $\frac{17}{100} =$ _0.17_

8. $\frac{8}{10} =$ _0.8_

9. $\frac{815}{1,000} =$ _0.815_

10. $\frac{2}{10} =$ _0.2_

Problem Solving

Processes &Practices **2** **Reason** Trudy is making a picture frame and
needs nails that measure 0.375 inch. At the hardware
store, nails are measured in fractions of an inch: $\frac{125}{1,000}$ inch,
$\frac{25}{100}$ inch, and $\frac{375}{1,000}$ inch. Which of these nails should she buy?

$\frac{375}{1000}$ in. because $0.375 = \frac{375}{1000}$

12. It rained 16 hundredths of an inch on Tuesday. Write
this amount as a decimal and a fraction.

0.16 $\frac{16}{100}$

Brain Builders

13. At Richardson Elementary, $\frac{35}{100}$ of the buses were late
because of a snowstorm. Write the fraction, the decimal,
and the word form of the buses that were *not* late.

$\frac{65}{100}$, 0.65 , Sixty five hundredths

14. **Test Practice** Which of the following does *not* represent
the number given by the model?

Ⓐ $\frac{4}{10}$

Ⓒ 0.4

Ⓑ *forty tenths*

Ⓓ *four tenths*

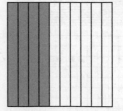

Check My Progress

Vocabulary Check

Choose the correct word(s) to complete each sentence.

decimal   **expanded form**   **period**   **standard form**

1. Each group of three digits on a place-value chart is called

   a(n) ___period___.

2. ___standard form___ is the usual or common way to write
   a number using digits.

3. A(n) ___decimal___ is a number that has a digit in the
   tenths place, hundredths place, and/or beyond.

4. A way of writing a number as the sum of the values of its digits

   is called ___expanded form___.

Concept Check

**Name the place of the highlighted digit. Then write the value
of the digit.**

5. 42,924,603

   ___2,000,000___

6. 953,187

   ___50,000___

7. Write 13,180,000 In expanded form.

   ___10,000,000 + 3,000,000 + 100,000 + 30,000___

8. Write 4,730,000 in word form.

   ___four million seven hundred thirty thousand___

Write <, >, or = in each ( ) to make a true sentence.

9. 84 (<) 90 **10.** 542 (>) 524 **11.** 925 (<) 1,024 **12.** 6,123 (<) 6,231

Shade the model. Then write each fraction in word form and as a decimal.

13. $\frac{1}{10}$

14. $\frac{85}{100}$

15. $\frac{39}{1,000}$

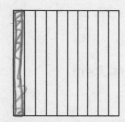

One tenth

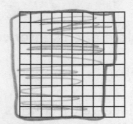

eighty five hundredths

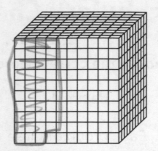

thirty nine thousandths

Problem Solving

16. The attendance at Friday's baseball game was 45,673. Sunday's game attendance was 45,761. Which game had a greater attendance?

45,761 Sundays game

Brain Builders

17. The shortest fish ever recorded is the dwarf goby, found in the Indo-Pacific. The female of this species is about $\frac{35}{100}$ inch long. Use a decimal to write the length of 2 females.

0.7 of two females

18. Test Practice Which decimal represents the shaded part of the figure?

(A) 0.0052

(C) 0.52

(B) 0.052

(D) 5.2

Name _____

Draw It

Use models to describe the relationship between the value of the digits in the decimal 0.77 and their place-value position.

1 Use models.

Shade the model to show 0.77. Write the decimal in word form and standard form.

Word form: *seventy- seven hundredths*

Standard form: 0.77

The place-value chart that you used for whole numbers can be extended to include decimals.

Write the decimal 0.77 in the place-value chart.

Tens	Ones	Tenths	Hundredths	Thousandths
◯	◯ .	7	7	◯

2 Describe the relationship.

The value of the digit 7 in the tenths place is 0.7 or __0.70__ *tenths*.

The value of the digit 7 in the hundredths place is 0.07 or

__0.070__ *hundredths*.

Use a calculator to find 0.7 ÷ 0.07. __10__

The value of the digit 7 in the tenths place is __10__ times as much as the value of the digit 7 in the hundredths place.

The value of the digit 7 in the hundredths place is $\frac{10}{1}$ times as much as the value of the digit 7 in the tenths place.

Try It

Describe the relationship between the value of the digits in the decimal 0.027 and their place-value position.

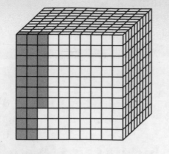

1 Use a place-value chart.

Write the decimal in the place-value chart.

Tens	Ones	Tenths	Hundredths	Thousandths
	0	0	2	7

2 Describe the relationship.

The digit in the thousandths place is ___7___.

It has a value of ___1000___.

If this digit were to move to the hundredths place,

it would have a value of ___100___.

If this digit were to move to the tenths place, it would have a value of ___10___.

The digit in each place has a value that is $\frac{1}{10}$ times as much as it has in the place to its left.

$\frac{1}{10} \times 100$

Talk About It

The decimal model shows 0.033. Use this decimal to answer Exercises 1–4.

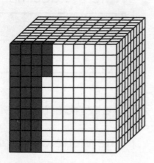

1. What is the value of the 3 in the hundredths place?

   _____30_____

2. What is the value of the 3 in the thousandths place?

   _____3_____

3. The digit in the hundredths place is how many times as much as the digit in the thousandths place?

   _____10_____ times

4. **Processes &Practices** **2** Use Number Sense How many times as great is the value of the digit in the thousandths place as in the hundredths place?

   _____10_____

Practice It

Use each model to write a decimal in standard form and word form.
Then complete each sentence.

5.

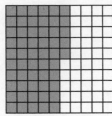

The value of the digit in the tenths place

is ___10___ times as much as the digit in the hundredths place.

fifty five hundredths   0.55

6.

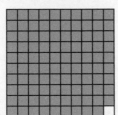

The value of the digit in the hundredths place

is ___10___ times as much as the digit in the tenths place.

Ninty nine hundredths   0.99

7.

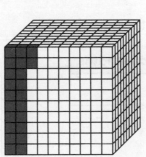

The value of the digit in the thousandths place

is ___10___ times as much as the digit in the hundredths place.

twenty two hundredths 0.22

8.

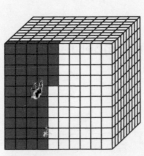

The value of the digit in the hundredths place

is ___10___ times as much as the digit in the thousandths place.

fourty four hundredths 0.44

9. Micah asked for directions to a local park. He was told to go 0.33 mile south to get to the park. The value of the digit in the tenths place is how many times as much as the value of the digit in the hundredths place?

10 times

10. **Processes &Practices** ➋ **Use Number Sense** In science class, Abigail's frog weighed 0.88 kilogram. What is the value of the digit in the tenths place?

0.80 Kilo.

11. **Processes &Practices** ➑ **Look for a Pattern** For the decimal 0.555, the value of the digit in the tenths place is how many times as much as the value of the digit in the thousandths place?

100 times

12. Look at the place-value chart. What happens to the value of the digit 2 as it moves places to the left from the thousandths place?

Tens	Ones	Tenths	Hundredths	Thousandths
	0 .	2	2	2

It began to grow the value.

Write About It

13. How are the place-value positions to either side of a particular number related?

The place values are related by being the same number and it is 10 times more.

Name ....

MY Homework

Homework Helper

Need help? connectED.mcgraw-hill.com

 The model shows a decimal. Write the decimal.

Word form: *eighty-eight hundredths*

Standard form: 0.88

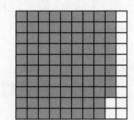

2 Write the decimal in the place-value chart.

Tens	Ones	Tenths	Hundredths	Thousandths
	0 •	8	8	

3 The value of the digit 8 in the tenths place is 0.8.

The value of the digit 8 in the hundredths place is 0.08.

The value of the digit 8 in the tenths place is 10 times
as much as the value of the digit 8 in the hundredths place.

The value of the digit 8 in the hundredths place is $\frac{1}{10}$ times
as much as the value of the digit 8 in the tenths place.

Practice

1. Use the model to write a decimal in standard form and
word form. Then complete the sentence.

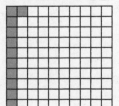

The value of the digit in the hundredths place is

_____10_____ times as much as the digit in the tenths place.

eleven hundredth 0. 11.

2. Use the model to write a decimal in standard form and word form. Then complete the sentence.

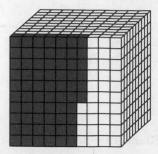

The value of the digit in the thousandths place

is _10_ times as much as the digit in the hundredths place.

Sixty six hundredth   0.066 $\frac{66}{1000}$

Problem Solving

3. Michael bought 0.44 pound of sliced turkey. What is the value of the digit in the hundredths place?

4

The value of the digit in the hundredths place is how many times as much as the value of the digit in the tenths place?

10   times

4. A small piece of metal weighs 0.77 gram. What is the value of the digit in the tenths place?

0.70   .70 grams

The digit in the tenths place is how many times as much as the value of the digit in the hundredths place?

10 times

5. Processes &Practices 2 **Use Number Sense** Write a decimal where the value of a digit is $\frac{1}{10}$ as much as a digit in another place.

0.77

Lesson 6
Place Value Through Thousandths

ESSENTIAL QUESTION
How does the position of a digit in a number relate to its value?

A decimal can be greater than one. For example, 1.5 is greater than one because there is a non-zero digit in the ones place.

Math in My World
Tools Tutor

Example 1

Five tree taps produce enough maple sap to make 1 gallon, or about 3.79 liters, of syrup. Read and write the number of liters in word form.

1 Write the number in the place-value chart.

Tens	Ones	Tenths	Hundredths	Thousandths
	3	7	9	

2 The place value of the last digit, 9, is ___hundredths___.

Use the word *and* for the decimal point.

Word form: ___three___ *and seventy-nine* ___hundredths___

Example 2

Circle the digit in the thousandths place. Then write the value of the digit.

0.247

1 The thousandths place is ___3___ places to the right of the decimal place.

2 The digit has a value of ___seven___ thousandths.

Example 3

Write *five and six hundred fourteen thousandths* in standard form and expanded form.

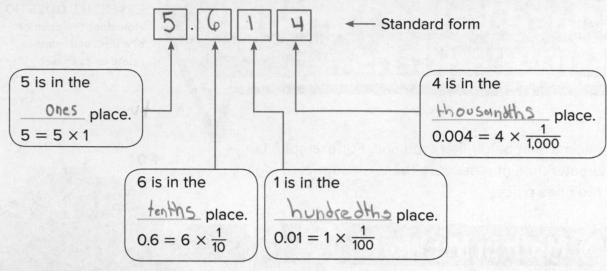

So, in expanded form, $5.614 = 5 \times 1 + \left(6 \times \frac{1}{10}\right) + \left(1 \times \frac{1}{100}\right) + \left(4 \times \frac{1}{1,000}\right)$.

Guided Practice

1. Circle the digit in the tenths place.   6.̲0̲14

2. Circle the digit in the hundredths place.   4.0③6

Write each number in standard form.

3. 5 and 87 hundredths

   5.87

4. $2 \times 10 + 6 \times 1 + \left(9 \times \frac{1}{10}\right) + \left(1 \times \frac{1}{100}\right) + \left(4 \times \frac{1}{1,000}\right)$

   26.014

5. Write 19.4 in expanded form. Then write in word form.

   $10 + 9 + \left(4 \times \frac{1}{10}\right)$ ninteen

   and four tenths

Talk MATH

Name the advantage of using 0.8 instead of $\frac{8}{10}$.

Independent Practice

Name the place of the highlighted digit. Then write the value of the digit.

6. 63.47

seven hundredths

7. 9.56

five tenths

8. 4.072

two thousandths

9. 81.453

four tenths

10. 1.608

eight thousandth

11. 7.017

One hundredths

Write each number in standard form.

12. *thirteen and nine tenths* $\underline{13.9}$

13. *fifty and six hundredths* $\underline{5.06}$

14. $1 \times 10 + 1 \times 1 + \left(9 \times \frac{1}{10}\right) + \left(2 \times \frac{1}{100}\right) + \left(3 \times \frac{1}{1000}\right)$ $\underline{11.923}$

15. $7 \times 10 + \left(1 \times \frac{1}{10}\right) + \left(5 \times \frac{1}{1000}\right)$ $\underline{70.105}$

16. *five and three thousandths* $\underline{5.003}$

17. $6 \times 10 + 4 \times 1 + \left(4 \times \frac{1}{10}\right) + \left(1 \times \frac{1}{100}\right) + \left(8 \times \frac{1}{1000}\right)$ $\underline{64.418}$

Write each number in expanded form. Then write in word form.

18. 0.917

$\left(9 \times \frac{1}{10}\right) + \left(1 \times \frac{1}{100}\right) + \left(7 \times \frac{1}{1000}\right)$ nine hundred seventeen thousandths

19. 69.409

$\left(6 \times 10\right) + \left(9 \times 1\right) + \left(4 \times \frac{1}{10}\right) + \left(9 \times \frac{1}{1000}\right)$ sixty nine and four hundred nine thousandths

Problem Solving

20. A baseball player had a batting average of 0.334 for the season. Write this number in expanded form.

$$\left(3 \times \frac{1}{10}\right) + \left(3 \times \frac{1}{100}\right) + \left(4 \times \frac{1}{1000}\right)$$

21. There were three and five hundredths inches of rain yesterday. Write this number in standard form.

3.05

22. An athlete completes a race in 55.72 seconds. How many times greater is the digit in the tens place than the digit in the ones place?

10 times

23. The table shows the amount of salt that remains when a cubic foot of water evaporates. Read each number that describes the amount of salt. Then write each number in words.

Salt Comparison	
Source of Water	**Amount of Salt**
Atlantic Ocean	2.2 pounds
Lake Michigan	0.01 pound

two and two tenths pounds

one hundredths pounds

Brain Builders

24. **Processes &Practices** **3** **Which One Doesn't Belong?** Circle the decimal that does not belong with the other three.

five and thirty-nine hundredths	5.39	$5 \times 1 + \left(3 \times \frac{1}{10}\right) + \left(9 \times \frac{1}{100}\right)$	5 and 39 tenths

25. **?** **Building on the Essential Question** How is place value used to read decimals?

like for example 5.39 in word form five and thirty nine hundredths the hundredths Part tell you that the 9 is in the hundredth and 3 tenths

Name ....

Homework Helper

Need help? connectED.mcgraw-hill.com

Write *six and seven hundred eighty-two thousandths* in standard form
and expanded form.

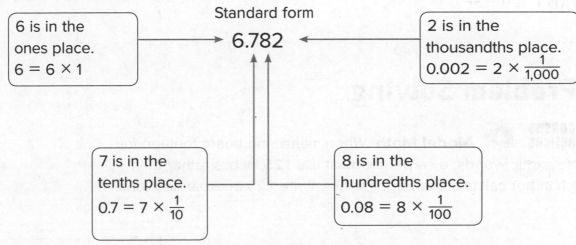

Standard form

6.782

6 is in the
ones place.
$6 = 6 \times 1$

2 is in the
thousandths place.
$0.002 = 2 \times \frac{1}{1,000}$

7 is in the
tenths place.
$0.7 = 7 \times \frac{1}{10}$

8 is in the
hundredths place.
$0.08 = 8 \times \frac{1}{100}$

So, in expanded form, $6.782 = 6 \times 1 + \left(7 \times \frac{1}{10}\right) + \left(8 \times \frac{1}{100}\right) + \left(2 \times \frac{1}{1,000}\right)$.

Practice

Name the place of the highlighted digit. Then write the value of the digit.

1. 35.052

    five hundredths

    5 /100

2. 5.654

    four thousandths

    4 /1000

3. 4.95

    nine tenths

    9 /10

Write each number in standard form.

4. *thirty-four and twelve hundredths* _____

5. $2 \times 10 + 4 \times 1 + \left(7 \times \frac{1}{10}\right) + \left(4 \times \frac{1}{100}\right) + \left(5 \times \frac{1}{1000}\right)$ _____

Write each number in expanded form. Then write in word form.

6. 23.5

_____

7. 164.38

_____

_____

8. 209.106

_____

_____

 **Problem Solving**

Processes &Practices **4** **Model Math** When measuring board footage for
9. some exotic woods, a carpenter must use 1.25 inches rather than
1 inch in her calculations for thickness. Write 1.25 in expanded form.

_____

Brain Builders

10. The summer camp Jessica attends is four hundred twenty-three
and four tenths miles from her home. A detour adds 10 miles to the
distance. How far does Jessica travel to camp? Write your answer
in standard form. Explain how you know.

_____

11. Test Practice Which statement is true regarding the value of the digit in the
tenths place of the decimal 19.993?

Ⓐ It is 10 times as great as the value of the digit in the ones place.

Ⓑ It is 10 times as great as the value of the digit in the thousandths place.

Ⓒ It is $\frac{1}{10}$ as great as the value of the digit in the ones place.

Ⓓ It is $\frac{1}{10}$ as great as the value of the digit in the tens place.

Name _____

Lesson 7
Compare Decimals

Math in My World
Watch | Tools | Tutor

Example 1

Tessa downloaded two songs onto her MP3 player. Which song is longer?

Song	Length (min)
1	3.6
2	3.8

One Way Use a number line.

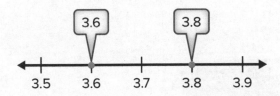

Numbers to the right are greater than numbers to the left.

Since 3.8 is to the right of 3.6, 3.8 ◯ 3.6.

Another Way Line up the decimal points.

1 Compare the digits in the greatest place.

The ones digits are the _____.

3.6
3.8

2 Continue comparing until the digits are different. In the tenths place, 8 ◯ 6.

So, 3.8 ◯ 3.6. Song _____ is longer.

Online Content at connectED.mcgraw-hill.com

Decimals that have the same value are **equivalent decimals**.

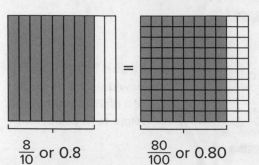

$\frac{8}{10}$ or 0.8 $\frac{80}{100}$ or 0.80

The models show you can annex, or place zeros, to the right of a decimal without changing its value.

Example 2

Write <, >, or = in the ⃝ below to make a true sentence.

8.69 ⃝ 8.6___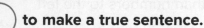

> Annex a zero to the right of 8.6 so that it has the same number of decimal places as 8.69.

Since 9 > 0 in the hundredths place, 8.69 ⃝ 8.6.

Guided Practice

Plot each decimal on the number line.

Write <, >, or = in each ⃝ to make a true sentence.

1. 0.5 ⃝ 0.7

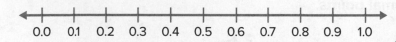

2. 4.40 ⃝ 4.44

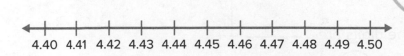

Talk MATH

Describe how you know if two decimals are equivalent.

Name _____

Independent Practice

Plot each decimal on the number line. Write <, >, or = in each ◯ to make a true sentence.

3. 4.4 ◯ 4.1

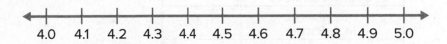

4. 0.37 ◯ 0.39

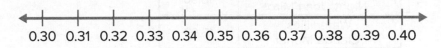

5. 0.57 ◯ 0.65

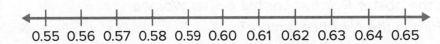

Write <, >, or = in each ◯ to make a true sentence.

6. 2.15 ◯ 2.150

7. 0.006 ◯ 0.1

8. 0.652 ◯ 0.647

9. 0.09 ◯ 0.001

10. 7.31 ◯ 7.304

11. 2.800 ◯ 2.8

12. 0.5 ◯ 0.7

13. 0.62 ◯ 0.26

14. 3.7 ◯ 3.70

Problem Solving

For Exercises 15–17, use the table that shows the cost of posters of famous works of art.

Poster Prices	
Poster	**Cost ($)**
From the Lake, No. 1, Georgia O'Keeffe	16.99
Relativity, M.C. Escher	11.49
Women and Bird in the Night, Joan Miro	18.98
Waterlillies, Claude Monet	15.99

15. Does the poster *Relativity* or the poster *Women and Bird in the Night* cost more?

_____

16. Which poster costs less: *From the Lake, No. 1* or *Waterlillies?*

_____

17. Which poster costs less than *Waterlillies?*

_____

Brain Builders

18. **Processes &Practices** 6 **Explain to a Friend** How many times greater is 46 than 0.46? Explain to a classmate.

_____

_____

19. **Building on the Essential Question** What are the similarities and differences between comparing whole numbers and comparing decimals?

_____

_____

_____

Name ....

Homework Helper

Need help? ⟋ connectED.mcgraw-hill.com

Compare 59.296 and 59.6.

1 Line up the decimal points. Annex zeros where necessary.

59.296
59.600

2 Compare the digits in the greatest place. The tens and ones digits are each the same.

Annex 2 zeros so that the numbers have the same number of decimal places.

3 Continue comparing until the digits are different. In the tenths place, 2 < 6.

So, 59.296 < 59.6.

Practice

Write <, >, or = in each ◯ to make a true sentence.

1. 3.976 ◯ 4.007

2. 89.001 ◯ 89.100

3. 126.698 ◯ 126.689

4. 5.05 ◯ 5.050

5. 9.087 ◯ 9.807

6. 3.674 ◯ 6.764

7. 0.256 ◯ 0.256

8. 2.7 ◯ 2.82

9. 6.030 ◯ 6.03

10. 7.89 ◯ 7.189

11. 12.54 ◯ 1.254

12. 0.981 ◯ 2.3

Problem Solving

13. In January, the average low temperature in Montreal, Quebec, Canada, is 5.2°F. The average low temperature in Cape Town, South Africa, is 60.3°F. Which city is warmer in January?

_____

14. In one year Detroit, Michigan, recorded 30.9 inches of snow and Chicago, Illinois, recorded 39.9 inches of snow. Which city had more snow?

_____

Brain Builders

15. **Processes &Practices** **6** **Explain to a Friend** George was weighed at the doctor's office. The scale read 67.20 pounds. The doctor wrote 67.2 pounds on George's chart. Did the doctor make a mistake? Explain to a friend.

_____

16. The two fastest times in the past 20 years for the girls' 200-meter run at Clarksville Elementary School are 27.97 seconds and 27.93 seconds. At yesterday's track meet, Claire ran 27.99 seconds. Was her time faster than either of the two fastest? Explain.

_____

Vocabulary Check

17. Write if the following statement is true or false.
Equivalent decimals are decimals that have the same value. _____

18. **Test Practice** Which of the following symbols make the statement below true?

98.546 〇 98.654

Ⓐ <   Ⓒ =

Ⓑ >   Ⓓ ≥

Name _____

Lesson 8
Order Whole Numbers and Decimals

ESSENTIAL QUESTION ❓
How does the position of a digit in a number relate to its value?

 Math in My World Watch Tools Tutor

Example 1

The table shows the cost to build three National Football League stadiums. Order the costs of the stadiums from *greatest* to *least*.

One Way Use place value.

1. Line up the numbers by their decimal points. →

   364.2
   430.0
   350.0

2. Compare the digits in the greatest place. _____ > 3

3. Compare the digits in the next greatest place.

   _____ > 5

Cost to Build (millions $)	
INVESCO Field Englewood, CO	364.2
Ford Field Detroit, MI	430.0
Qwest Field Seattle, WA	350.0

Another Way Use a number line.

Place dots on the number line to represent the approximate locations of the decimals.

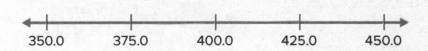

350.0   375.0   400.0   425.0   450.0

So, the costs, in millions of dollars, from greatest to least are _____,

_____, and _____.

Example 2

Four trees in a state forest had heights of 22.65, 23.8, 22, and 23.25 feet. Order the heights from _least_ to _greatest_.

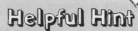

1 Line up the numbers by their decimal points.

2 Annex zeros so that all numbers have the same final place value.

3 Compare the digits using place value.

22.65
23.80
22.00
23.25

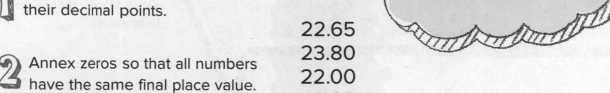
Helpful Hint

To annex a zero means to add a zero at the end of a number.

The least number is _____.

The greatest number is _____.

The heights, in feet, from least to greatest are _____, _____, _____,

and _____.

Guided Practice

Order each set of numbers from _least_ to _greatest_.

1. weight in kilograms of a dog: 56.7, 64.3, 59.0, 64.5

_____

2. rainfall in inches: 0.76, 0.09, 0.63, 0.24

_____

3. height of flowers in inches: 8.9, 8.59, 8.705, 8.05

_____

4. The lengths of insects in centimeters are 1.35, 0.9, 1.48, and 1.8. Order the sizes of the insects from greatest to least.

_____

Talk MATH

Discuss different steps that make ordering numbers easier.

Independent Practice

Order each set of numbers from *least* to *greatest*.

5. cost of cellphones: $98.75, $114.99, $105.99

_____

6. temperatures in °F: 106.3, 99.8, 101.1, 110.5

_____

7. distance in light years: 4.2, 6.0, 4.3, 7.7

_____

8. heights of buildings in meters: 419.7, 346.5, 178.3, 527.3

_____

9. kilometers ran: 4.9, 3.7, 3.4, 4.2

_____

Order each set of numbers from *greatest* to *least*.

10. cost of snacks: $2.43, $2.34, $2.05, $2.18

_____

11. masses of bottles in grams: 9.14, 7.99, 9.02, 8.95, 8.91

_____

12. race times in seconds: 43.789, 67.543, 86.347, 78.432, 34.678

_____

13. heights of trees in meters: 9.8, 10, 10.2, 9.6, 11

_____

14. weights of dogs in pounds: 25.4, 26.2, 26, 25.8, 27

_____

Problem Solving

Processes &Practices **3** **Draw a Conclusion** The table shows facts about snakes common to the United States.

Snake	Average Adult Body Length (cm)	Average Baby Body Length (cm)
Copperhead	63.5	27.9
Western Cottonmouth	91.25	21.5
Timber Rattlesnake	121.6	29.5
Queen Snake	61	15.2

15. List the average baby body lengths, in centimeters, from *least* to *greatest*.

16. Write the names of the snakes in order from *greatest* to *least* average adult body length.

17. The average length of an adult Eastern Coachwhip snake is 152.4 centimeters. Write a sentence comparing its length to the length of the other snakes listed in the table.

Brain Builders

Processes &Practices **1** **Make Sense of Problems** Write an ordered list of five **18.** numbers whose values are between 50.98 and 51.6. Tell whether your list is from least to greatest or greatest to least.

19. **Building on the Essential Question** How does comparing numbers help to order numbers?

MY Homework

Homework Helper

Need help? connectED.mcgraw-hill.com

Order the set of numbers 9.275, 8.950, and 9.375 from *least* to *greatest*.

Use place value.

1 Line up the numbers by
their decimal points.

2 Compare the digits in the
greatest place. 9 > 8

9.275
8.950
9.375

3 Compare the digits in the
next place. 3 > 2

So, the order from least to greatest is 8.950, 9.275, and 9.375.

Practice

Order each set of numbers from *least* to *greatest*.

1. 17.639, 3.828, 45.947

_____

2. 890.409, 890.904, 809.904

_____

Order each set of numbers from *greatest* to *least*.

3. 2.654, 2.564, 2.056, 2.465

_____

4. 1.11, 0.111, 1.01, 1.001

5. **Processes &Practices** ➋ **Use Number Sense** The table shows the heights of four students. Arrange the students in order from shortest to tallest.

Student Heights	
Name	**Height (in.)**
Kim	56.03
Alexa	56.3
Roy	56.14
Tom	57.1

_____

⚙ Brain Builders

6. Lauren spent $3.26 for lunch on Tuesday. She spent $3.56 on Wednesday and $3.29 on Thursday. Order the prices from greatest to least on the line below. Draw a number line in the space to support your answer.

_____

7. The four fastest times in a race were 27.08 seconds, 27.88 seconds, 27.8 seconds, and 26.78 seconds. Between which two times does 27.84 seconds belong?

_____

8. **Test Practice** Four boxes to be mailed are weighed at the post office. Box A weighs 8.25 pounds, Box B weighs 8.2 pounds, Box C weighs 8.225 pounds, and Box D weighs 8.05 pounds. Which box is the heaviest?

Ⓐ Box A   Ⓒ Box C

Ⓑ Box B   Ⓓ Box D

Lesson 9
Problem-Solving Investigation
STRATEGY: Use the Four-Step Plan

ESSENTIAL QUESTION
How does the position of a digit in a number relate to its value?

Learn the Strategy

Victor spent **$61** on some sandpaper for his model cars. He bought 2 packages of the smallest-grain sandpaper and spent the rest on the largest-grain sandpaper. How many packages of the largest-grain sandpaper did he buy?

Size of Grain (cm)	Cost per Package ($)
0.003	13
0.011	7
0.001	20

1 Understand

What facts do you know?

A total of _____ was spent. _____ packages of the smallest-grain sandpaper were purchased.

What do you need to find?

the number of packages of the largest grain sandpaper he bought

2 Plan

To solve this problem, I can work backward.

3 Solve

The smallest size is _____ centimeter. Victor spent 2 × _____ or _____.

Subtract to find the remaining amount spent: $61 − _____ = _____.

The largest size is 0.011 centimeter. Each package costs $7. Divide.

Victor bought $21 ÷ $7, or _____ packages of the largest grain sandpaper.

4 Check

Does your answer make sense? Explain.

_____

Luisa bought a roll of ribbon. She used 34 inches on each of two gifts. Then she used 13 inches on a scrapbook page. There are 39 inches left. How many inches did she start with?

 Understand

What facts do you know?

_____

_____

What do you need to find?

_____

 **Plan**

_____

_____

 **Solve**

 **Check**

Is your answer reasonable? Explain.

_____

Apply the Strategy

Solve each problem using the four-step plan.

Recipe	Ounces of Butter
Pie	4
Cookie	8
Pasta	6

1. The table shows the number of ounces of butter Marti used in different recipes. She has 6 ounces of butter left. How many ounces of butter did she have at the beginning? Is your answer reasonable? Explain.

_____

2. At the end of their 3-day vacation, the Palmers traveled a total of 530 miles. On the third day, they drove 75 miles. On the second day, they drove 320 miles. How many miles did they drive the first day? Is your answer reasonable? Explain.

_____

Brain Builders

Processes &Practices **1** **Keep Trying** You divide a number by 3, add 6, then subtract 7. The result is 4. What is the number? Explain.

_____

4. Mr. Toshio lent out 11 rulers at the beginning of class, collected 4 rulers in the middle of class, and gave out 7 at the end of class. He had 18 at the end of the day. How many rulers did he start with? What is the first step to solving this problem?

_____

5. The Math Club is selling gift wrap for a fundraiser. They sold all 45 rolls of solid wrapping paper at $4 each and rolls of patterned wrapping paper at $5 each. If they made $265, how many rolls of patterned wrapping paper did they sell?

_____

Review the Strategies

Use any strategy to solve each problem.

- Use the four-step plan.
- Make a table.
- Act it out.

Processes & Practices ➊ **Make a Plan** Ming-Li spent $15 at the movies. She then earned $30 babysitting. She spent $12 at the bookstore. She now has $18 left. How much money did Ming-Li have to begin with?

6.

_____

7. Mr. Jenkins bought pavers for some landscaping projects. He used 120 for a small patio, 86 for the border of a flower bed, and 70 for a wall. He has 24 pavers left. How many pavers did Mr. Jenkins buy?

_____

8. The table shows the number of candy bars the cheerleaders sold each week. They have 9 candy bars left. How many candy bars did they have to sell to begin with?

Week	Candy Bars Sold
1	117
2	130
3	83

_____

9. You multiply a number by 3, subtract 6 and then add 2. The result is 20. What is the number?

_____

10. Dimitrius returned 5 library books last week, returned 3 books this week, and then checked out 8 more books. He now has 12 library books. How many books did he have before last week?

_____

MY Homework

Homework Helper

Need help? connectED.mcgraw-hill.com

Dana walks every day. She walked 3 miles on Tuesday, 5 miles on Wednesday, and 8 miles on Thursday. After Thursday's walk, she had walked a total of 21 miles for the week. How many miles did she walk on Monday?

1 Understand

What facts do you know?

Dana walked 3 miles on Tuesday, 5 miles on Wednesday, and 8 miles on Thursday. Dana walked a total of 21 miles from Monday to Thursday.

What do you need to find?

the number of miles Dana walked on Monday

2 Plan

I can solve the problem by adding 3, 5, and 8 and then subtracting the sum from 21.

3 Solve

$$21 - (3 + 5 + 8) = 21 - 16$$
$$= 5 \text{ miles}$$

So, Dana walked 5 miles on Monday.

4 Check

Does your answer make sense? Explain.

Yes. $5 + 3 + 5 + 8 = 21$

Problem Solving

Solve each problem using the four-step plan.

Day	Vehicles
Fri.	27
Sat.	?
Sun.	34

1. The table shows the number of vehicles washed at a car wash fundraiser over the weekend. If there were a total of 94 vehicles, how many were washed on Saturday? Explain.

2. The volleyball team sold 16 items on the first day of a bake sale, 28 items the second day, and 12 items the last day. There were 4 items left that had not been purchased. How many total items were for sale at the bake sale? Explain how you know.

Brain Builders

3. Ricardo lost 6 golf balls while playing yesterday. He bought a box of 12 golf balls, then lost 4 on the course today. He now has 18 golf balls. How many golf balls did Ricardo have to begin with? What is the first step to solving this problem?

4. The distance between Cincinnati, Ohio, and Charlotte, North Carolina, is about 336 miles. The distance between Cincinnati and Chicago, Illinois, is about 247 miles. If Perry drove from Charlotte to Chicago by way of Cincinnati, find the distance he drove. Show your work.

5. **Processes &Practices** ➊ **Plan Your Solution** Adison earned $25 mowing her neighbor's lawn. Then she loaned her friend $18, and got $50 from her grandmother for her birthday. She now has $86. How much money did Adison have to begin with? Is your answer reasonable? Explain.

Need more practice? Download Extra Practice at **connectED.mcgraw-hill.com**

Review

Vocabulary Check

Choose the correct word(s) to complete each sentence.

| decimal | decimal point | equivalent decimals | expanded form |
| period | place value | place-value chart | standard form |

1. Decimals that have the same value are _____.

2. _____ is a system for writing numbers. In this system, the position of a digit determines its value.

3. The usual or common way to write a number is called

    _____.

4. The way of writing a number as the sum of the values

    of its digits is called _____.

5. A _____ is a number that has a digit in the tenths place, hundredths place, and/or beyond.

6. The _____ is a period separating the ones and the tenths in a decimal number.

7. A _____ is a chart that shows the value of the digits in a number.

Concept Check

Name the place of the highlighted digit. Then write the value of the digit.

8. 195,489

_____

9. 6,720,341

_____

Write each number in standard form.

10. 94 million, 237 thousand, 108 _____

11. 8 × 1,000,000 + 5 × 10,000 + 2 × 1,000 + 6 × 100 _____

Shade the model to show each fraction. Write each fraction as a decimal.

12. $\frac{19}{100}$ 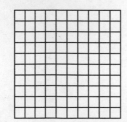 _____

13. $\frac{8}{10}$ 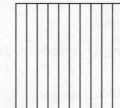 _____

Write each number in standard form and expanded form.

14. *five and nine tenths*

_____

15. *seven hundred twelve thousandths*

_____

Write <, >, or = in each ( ) to make a true sentence.

16. 14,589 ( ) 14,985

17. 506,789 ( ) 505,789

18. 8,913 ( ) 8,931

19. 0.49 ( ) 0.71

20. 9.02 ( ) 9.020

21. 0.843 ( ) 0.846

22. Order the set of numbers from *least* to *greatest*.

13.84, 13.097, 12.655, 13.6

_____

 # Problem Solving

23. One cup is equal to $\frac{5}{10}$ pint. Write this fraction as a decimal.

_____

24. Charles is moving from Springfield, which has 482,653 people, to Greenville, which has 362,987 people. Is he moving to a city with a greater or smaller population? Explain.

_____

 **Brain Builders**

25. The population of Texas is about 27,500,000. The population of Florida is about 20,300,000. Write the population of Texas and Florida combined in word form.

_____

26. Test Practice Which number results in a true sentence in 0.475 < _____?

Ⓐ 0.473

Ⓒ 0.475

Ⓑ 0.474

Ⓓ 0.476

Reflect

Use what you learned about place value to complete the graphic organizer.

Example with Whole Numbers

ESSENTIAL QUESTION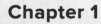

How does the position of a digit in a number relate to its value?

Example with Decimals

Now reflect on the ESSENTIAL QUESTION Write your answer below.

_____

_____

_____

Performance Task

Setting Goals

A factory produces components for game consoles. The new manager wants to set a goal for how many components will be produced in the upcoming year.

Show all your work to receive full credit.

Part A

The factory has existed for seven years. The chart below gives the number of components produced by the factory each year.

Year	Components
1	2,659,051
2	2,500,197
3	2,834,180
4	2,384,201
5	2,799,125
6	2,679,051
7	2,384,198

The factory manager needs to put the data in order so that he can make a decision on next year's goal. Order the data from least to greatest.

Part B

The factory manager asks his assistant manager to give input for the production goal. The assistant manager suggests 2,300,000 components. Explain why this goal may not be appropriate.

Part C

While the manager is tempted to set a new record producing more components than ever before, he knows that people are not buying as many components as they used to. He does not want to make more components than can be sold. He decides to set the goal of producing the third highest number of components in company history. Suggest a goal for the factory manager.

Part D

In researching the company financial reports, the factory manager discovers that the factory must produce at least 2,700,000 components in a year in order to make a profit. Does your goal from **Part C** meet this requirement? If so, explain why. If not, offer the factory manager a new goal that meets both requirements.

Part E

The factory manager's supervisor indicates that it is absolutely essential that the total number of components sold in years 6, 7, and 8 (the new year) be at least 7,709,000. Explain why the goal you gave the factory manager in **Part D** will also meet this new requirement.

Chapter 2 Multiply Whole Numbers

Taking Care of My Pets

ESSENTIAL QUESTION

What strategies can be used to multiply whole numbers?

Watch

Watch a video!

Name ....

MY Chapter Project

About How Much?

1. On a separate sheet of paper, keep a food diary and estimate how much of each food you eat for one day. Your diary should include breakfast, lunch, dinner, and snacks.

2. Use the Internet or nutritional resources to compare your food intake with actual serving sizes. Create a chart in the space at the right that compares your estimated intake of each food to what nutrition guides list as the recommended daily allowance.

3. Choose a food item from your chart. Estimate how much of that food would be eaten in

2 days _____   1 week _____

1 month _____   3 months _____

6 months _____   1 year _____

4. Choose 1 of your estimates above. Use the space below to show the steps you used to find your estimates.

Name
....

Am I Ready?

Write all of the factors of each number.

1. 8 ____ , ____ , ____ , ____

2. 11 ____ , ____

3. 6 ____ , ____ , ____ , ____

4. 15 ____ , ____ , ____ , ____

5. 32 ____ , ____ , ____ , ____ , ____ , ____

6. 10 ____ , ____ , ____ , ____

Write each repeated addition sentence as a multiplication sentence.

7. $5 + 5 + 5 + 5 = 20$

_____

8. $8 + 8 + 8 = 24$

_____

9. $21 + 21 = 42$

_____

10. $6 + 6 + 6 + 6 + 6 = 30$

_____

11. $13 + 13 + 13 = 39$

_____

12. $7 + 7 + 7 + 7 + 7 + 7 = 42$

_____

Multiply.

13. $6 \times 3 =$ _____

14. $1 \times 8 =$ _____

15. $7 \times 8 =$ _____

16. $4 \times 10 =$ _____

17. Nikki purchased three used books at a garage sale for $5 each. Find the total cost for all three books.

_____

How Did I Do? Shade the boxes to show the problems you answered correctly.

1	2	3	4	5	6	7	8	9	10	11	12	13	14	15	16	17

Online Content at connectED.mcgraw-hill.com

MY Math Words

Vocab
abc

Review Vocabulary

composite numbers   prime numbers

Making Connections

Complete the chart below using the review words.

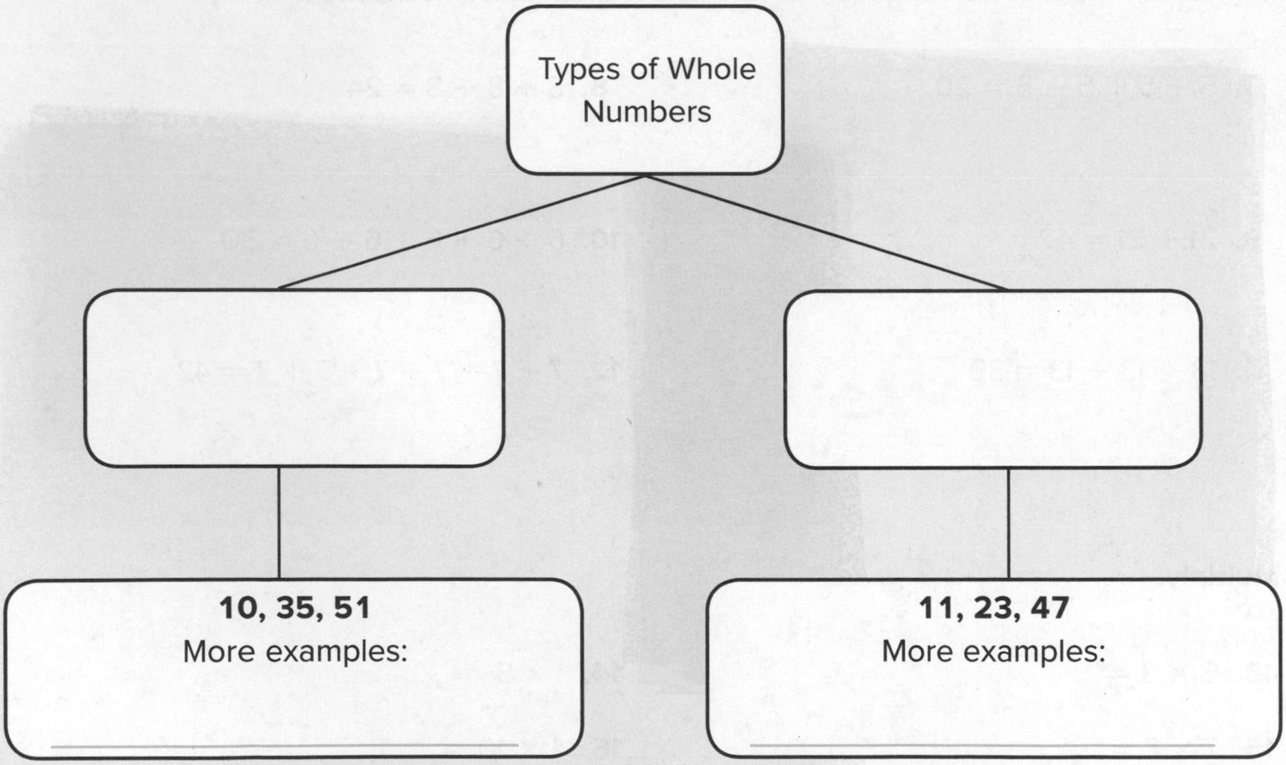

Types of Whole
Numbers

[   ]   [   ]

10, 35, 51
More examples:

_____

11, 23, 47
More examples:

_____

Explain how you decided to categorize each set of numbers.

_____

_____

_____

MY Vocabulary Cards

Processes & Practices

Lesson 2–3

base

$$3^3 = 3 \times 3 \times 3 = 27$$

$$5^4 = 5 \times 5 \times 5 \times 5 = 625$$

Lesson 2–8

compatible numbers

$$42 \times 7 = 294$$
$$\underbrace{40 \times 7 = 280}$$

compatible numbers

Lesson 2–3

cubed

$$3^3 = 3 \times 3 \times 3 = 27$$

Lesson 2–6

Distributive Property

$$4 \times (2 + 7) = (4 \times 2) + (4 \times 7)$$

Lesson 2–3

exponent

$$5^4 = 5 \times 5 \times 5 \times 5 = 625$$

Lesson 2–3

power

$$3^4 = 3 \times 3 \times 3 \times 3 = 81$$

81 is a power of 3.

Lesson 2–4

power of 10

$$10^4 = 10 \times 10 \times 10 \times 10$$
$$= 10,000$$

Lesson 2–1

prime factorization

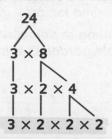

Ideas for Use

- Work with a partner to name the parts of speech of each word. Consult a dictionary to check your answers.

- Write a tally mark on each card every time you read the word in this chapter or use it in your writing. Try to use at least 2 to 3 tally marks for each card.

Numbers in a problem that are easy to work with mentally.

How can the meaning of *compatible* help you remember this definition?

_____

_____

In a power, the number used as a factor.

Consult a dictionary to find another math-related definition of *base*. Describe that meaning in your own words.

_____

_____

To multiply a sum by a number, multiply each addend by the number, and add the products.

How can the meaning of the verb *distribute* help you remember this property?

_____

_____

A number raised to the third power.

Consult a dictionary to describe how *cube* is used in geometry. Write that meaning.

_____

_____

A number obtained by raising a base number to an exponent.

Add a suffix to *power* to form a new word.

_____

_____

In a power, the number of times the base is used as a factor.

What suffix can you add to *exponent* to make a word meaning "growing or increasing very rapidly"? Name the word.

_____

_____

A way of expressing a composite number as a product of its prime factors.

Identify the meaning of *prime* in this sentence: *Maria's vegetable garden is in its prime in mid-July.*

_____

A number like 10, 100, 1,000 and so on. It is the result of using only 10 as a factor.

Is the number 30 a power of 10? Explain.

_____

_____

MY Vocabulary Cards

property

Commutative Property $25 \times 47 = 47 \times 25$

Identity Property $25 \times 0 = 0$

Distributive Property $5 \times (2 + 7) = (5 \times 2) + (5 \times 7)$

squared

$$25^2 = 25 \times 25 = 625$$

Ideas for Use

- Use the back of the card to write or draw examples to help you answer the question.

- Use the blank cards to write notes about important concepts in this chapter, such as how powers of 10 apply to multiplication patterns.

A number raised to the second power.

Write a sentence using *square* as a noun.

_____

_____

A rule in mathematics that can be applied to all numbers.

Write a sentence using *property* in a non-mathematical way.

_____

_____

✂

2 × 24

3 × 16

4 × 12

 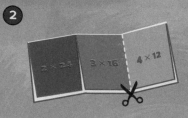
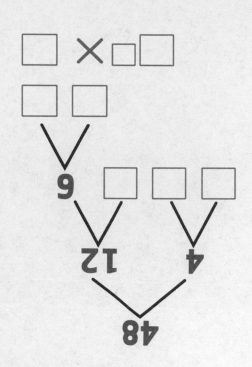

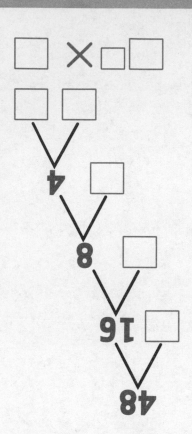

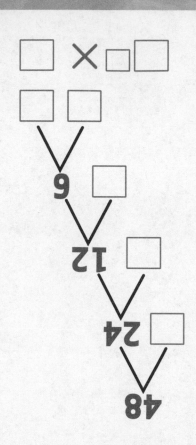

Name ....

Lesson 1
Prime Factorization

ESSENTIAL QUESTION
What strategies can be used to multiply whole numbers?

You can write every composite number as a product of prime factors. This is called the **prime factorization** of a number. A factor tree is a diagram that shows the prime factorization of a composite number.

 Math in My World Watch ▶ Tutor 💬

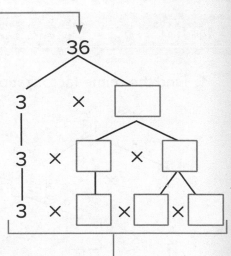

Example 1

Mr. Dempsey surveyed his class and found that his students have a total of 36 pets. Find the prime factorization of 36.

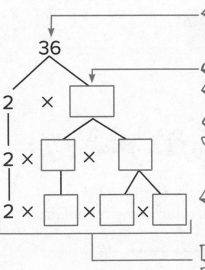

1 Write the number to be factored at the top.

2 Choose any pair of whole number factors of 36.

3 Continue to factor any number that is not prime.

4 Except for the order, the prime factors are the same.

5 Write the prime factors in order from least to greatest.

The prime factorization of 36 is ____ × ____ × ____ × ____.

Check Work backward. Multiply all the prime factors in order from left to right. Then compare your product with the composite number.

2 × 2 × 3 × 3 = _____

Copyright © McGraw-Hill Education   G.K. & Vikki Hart/Photodisc/Getty Images

Online Content at 🔗 **connectED.mcgraw-hill.com**

Lesson 1 **81**

Example 2

Find the prime factorization of 24.

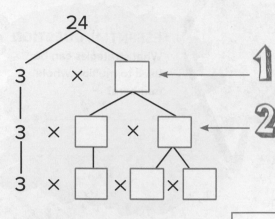

1 You can choose any pair of whole number factors, such as 3 × 8, 4 × 6, or 2 × 12.

2 Continue to factor any number that is not prime.

3 Write the prime factors in order from least to greatest.

The prime factorization of 24 is ____ × ____ × ____ × ____ .

Check

Work backward. Multiply all the prime factors in order from left to right. Then compare your product with the composite number.

2 × 2 × 2 × 3 = _____

Guided Practice

1. Find the prime factorization of 16.

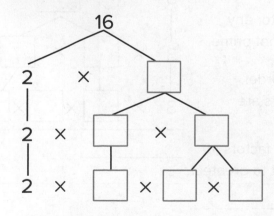

The prime factorization of 16 is

_____ × _____ × _____ × _____ .

Talk MATH

What are the first ten prime numbers?

Independent Practice

Find the prime factorization of each number.

2. 63 = _____

3. 18 = _____

4. 40 = _____

5. 75 = _____

6. 27 = _____

7. 32 = _____

8. 49 = _____

9. 44 = _____

Processes &Practices **Understand Symbols** Find the missing number.

10. $104 = 2 \times 2 \times \blacksquare \times 13$

$\blacksquare = $ _____

11. $55 = \blacksquare \times 11$

$\blacksquare = $ _____

12. $77 = 7 \times \blacksquare$

$\blacksquare = $ _____

Problem Solving

Use the table for Exercises 13–16 that shows the average weights of popular dog breeds.

Breed	Weight (lb)	Prime Factorization
Cocker Spaniel	20	
German Shepherd	81	
Labrador Retriever	67	
Beagle	25	
Golden Retriever	70	
Siberian Husky	50	

13. Complete the table.

14. Which weight(s) have a prime factorization of exactly three factors?

_____

15. Which weight(s) have a prime factorization with factors that are all the same number?

_____

Brain Builders

16. Which breed(s) have weights that are prime numbers?

_____

A Lakeland Terrier weighs less than 20 pounds. List a new prime number and a new composite number that could be the average weight of the breed. Explain how you know.

_____

17. **Processes &Practices** ▶7 **Identify Structure** Find the prime factorization of 2,800. Draw a factor tree and explain each step to a classmate.

_____

_____

_____

_____

18. **Building on the Essential Question** How do factor trees help you find the prime factorization of a number?

_____

Lesson 1

Prime Factorization

Homework Helper

Need help? ➣ **connectED.mcgraw-hill.com**

Find the prime factorization of 42.

1 You can choose any pair of whole number factors, such as 2 × 21, 3 × 14, or 6 × 7.

2 Continue to factor any number that is not prime.

3 Write the prime factors in order from least to greatest.

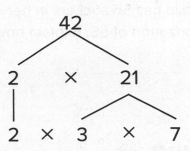

The prime factorization of 42 is 2 × 3 × 7.

Practice

Find the prime factorization of each number.

1. 50 = _____

2. 81 = _____

3. 65 = _____

4. 28 = _____

Problem Solving

5. Britney scored an 85 on her last math test. Write the prime factorization of 85.

_____

Brain Builders

6. Priscilla has 56 stickers in her collection. Write the prime factorization of 56. Explain how you know your numbers are prime.

_____

_____

7. Processes &Practices 3 **Find the Error** Lainey wrote the prime factorization of 60 as 2 × 5 × 6. Is she correct? If not, what is the prime factorization of 60? Explain.

_____

_____

Vocabulary Check

Fill in each blank with the correct term or number to complete the sentence.

8. _____ numbers can be written as a product of _____ factors. This is called the prime factorization of a number.

9. Test Practice Josiah had a pot belly pig as a pet that weighed 46 pounds. What is the prime factorization of 46?

(A) 2 × 23   (C) 2 × 2 × 13

(B) 2 × 2 × 11   (D) 3 × 23

Lesson 2
Hands On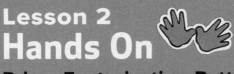
Prime Factorization Patterns

ESSENTIAL QUESTION
What strategies can be used to multiply whole numbers?

Build It

You can make a pattern using paper and a hole punch. By folding the paper, hole punching it, and counting the holes, you can discover a pattern.

1 Fold a piece of paper in half and make one hole punch. Open the paper.

How many holes are in the paper? _____

Find the prime factorization for the number of holes. _____

2 Fold another piece of paper in half twice. Make one hole punch.

Unfold the paper. How many holes are in the paper? _____

What is the prime factorization for the number of

holes? _____ × _____

3 Complete the table for one, two, and three folds.

Number of Folds	Number of Holes	Prime Factorization
1		
2		
3		

Online Content at **connectED.mcgraw-hill.com**

4 What pattern do you notice between the number of factors in each prime factorization and the number of folds?

_____

_____

_____

5 Using the pattern you found in Step 4, complete the table for four and five folds.

Number of Folds	Number of Holes	Prime Factorization
1	2	2
2	4	2 × 2
3	8	2 × 2 × 2
4		
5		

Talk About It

1. Which prime number did you record in each prime factorization?

_____

2. How many holes will you make if you fold the paper eight times? Write the prime factorization of that number.

_____

Processes & Practices ➡ **1** **Make Sense of Problems** How can you check
3. that you have the correct prime factorization?

_____

_____

_____

Practice It

4. Use paper and a hole punch to complete the table below. Start by folding a piece of paper in half and making 3 holes. Use a new piece of paper each time you increase the number of folds.

Number of Folds	Number of Holes	Prime Factorization
1	6	2 × 3
2	12	2 × 2 × 3
3		
4		
5		

Find a pattern to complete the tables in Exercises 5–7.

5.

Number of Folds	Number of Holes	Prime Factorization
1	10	2 × 5
2	20	2 × 2 × 5
3	40	
4		
5		

6.

Number of Folds	Number of Holes	Prime Factorization
1	14	2 × 7
2	28	2 × 2 × 7
3	56	
4		
5		

7.

Number of Folds	Number of Holes	Prime Factorization
1	18	2 × 3 × 3
2	36	2 × 2 × 3 × 3
3	72	
4		
5		

Apply It

Use the information below to solve Exercises 8–11. There is one skin cell used in a science lab. Each day, the skin cell will split into two cells. The next day the cell then splits into two cells again.

Number of Days Passed	Number of Cells
1	2
2	4
3	8
4	16
5	32

8. After several splits, there are 64 cells. How many days have passed?

9. How many skin cells will there be after 8 days have passed?

10. How many days would need to pass before there were over 2,000 cells?

11. **Processes &Practices** 1 **Make a Plan** After 15 days, there were 32,768 cells. After how many days were there 16,384 cells?

Write About It

12. How can I use patterns to describe relationships?

Name _____

Homework Helper

Need help? ➚ **connectED.mcgraw-hill.com**

A design with equilateral triangles is shown. The triangle is divided into four equal-sized, smaller triangles as shown. Then each of the four triangles is divided into four equal-sized, smaller triangles. If the pattern continues, how many triangles will there be in Figure 3?

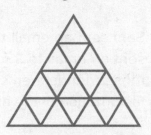

Figure 1

In Figure 1, there are 4 resulting triangles. In Figure 2, there are 16 triangles.

How many triangles will be in Figure 3?

The table shows the Figure numbers, the number of triangles formed, and the prime factorization of the number of triangles.

Figure 2

Figure Number	Number of Triangles Formed	Prime Factorization
1	4	2×2
2	16	$2 \times 2 \times 2 \times 2$
3	64	$2 \times 2 \times 2 \times 2 \times 2 \times 2$

By following the pattern, there are 64 triangles formed in Figure 3.

Practice

1. Complete the table for Figures 4 and 5 if the above pattern continues.

Figure Number	Number of Triangles Formed	Prime Factorization
4		
5		

2. A population of rabbits triples every month. The population starts with two rabbits. How many rabbits are there after three months?

_____

3. Three friends each create 4 bags of starter bread dough. After ten days, each of those four bags is then divided into four more bags of dough. How many days pass before 192 bags of dough are created?

_____

4. Sam sent an email to 3 friends on Monday. Each of the friends then sent an email to 3 friends on Tuesday. On Wednesday, each of those friends then sent an email to 3 friends. Write the prime factorization of the number of emails that were sent on Wednesday.

_____

5. **Processes &Practices** **8** **Look for a Pattern** Annie opened a savings account and deposited $10. If the balance in her account doubles each month, what is the account balance after 4 months?

_____

6. Elyse folded a piece of paper in half 3 times. She then punched 3 holes in the paper. How many holes are in the paper when she unfolds it?

_____

Lesson 3
Powers and Exponents

ESSENTIAL QUESTION ❓
What strategies can be used to multiply whole numbers?

A product of identical factors can be written using an exponent and a base. The **base** is the number used as a factor. The **exponent** indicates how many times the base is used as a factor.

 Math in My World ▶ Watch 💬 Tutor

Example 1

The number of Calories in six pancakes can be written as 10^3. Write 10^3 as a product of the same factor. Then find the value.

$$\underbrace{10 \times 10 \times 10}_{\text{3 factors}} = 10^{\overset{\text{exponent}}{3}}_{\text{base}}$$

$10 \times 10 \times 10 =$ _____

Six pancakes have _____ Calories.

Numbers expressed using exponents are called **powers.** Numbers raised to the second or third power have special names.

Powers	Words
2^5	2 to the fifth power
3^2	3 to the second power or 3 **squared**
10^3	10 to the third power or 10 **cubed**

Example 2

Write 3 × 3 × 3 × 3 using an exponent.

The base is _____ . Since 3 is used as a factor _____ times,

the exponent is _____ .

Write as a power. $3 \times 3 \times 3 \times 3 =$ _____

Example 3

Write the prime factorization of 72 using exponents.

1 Complete the factor tree.

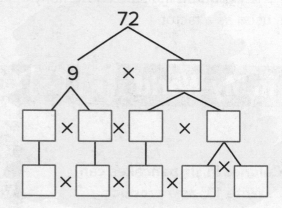

2 Order the factors from least to greatest.

_____ × _____ × _____ × _____ × _____

3 Write products of identical factors using exponents.

$2^{\square} \times 3^{\square}$

So, 72 = _____ × _____ .

Talk MATH

Explain how a factor tree helps you to write the prime factorization of a number using exponents.

Guided Practice

1. Write 4 × 4 × 4 × 4 using an exponent.

   The base is _____ . Since 4

   is used as a factor _____

   times, the exponent is _____ .

   $4 \times 4 \times 4 \times 4 =$ _____

Name _____

Independent Practice

Write each product using an exponent.

2. $10 \times 10 =$ _____

3. $8 \times 8 \times 8 \times 8 =$ _____

4. $3 \times 3 \times 3 \times 3 \times 3 \times 3 =$ _____

5. $5 \times 5 \times 5 \times 5 \times 5 =$ _____

6. $9 \times 9 \times 9 \times 9 =$ _____

7. $1 \times 1 \times 1 \times 1 \times 1 =$ _____

Write each power as a product of the same factor. Then find the value.

8. $10^4 =$ _____

9. $3^2 =$ _____

10. $9^3 =$ _____

11. $6^5 =$ _____

Write the prime factorization of each number using exponents.

12. $25 =$ _____

13. $56 =$ _____

14. $68 =$ _____

15. $88 =$ _____

Problem Solving

16. To find the amount of space a cube-shaped bird cage occupies, find the *cube* of the measure of one edge of the bird cage. Express the amount of space occupied by the bird cage shown as a power. Then find the amount in cubic units.

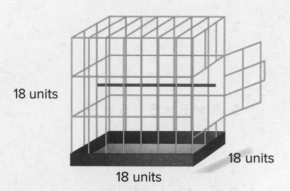

18 units

18 units

18 units

17. A single tusk that weighed just over 2^8 pounds from an African elephant is the largest tooth ever recorded from any modern animal. About how many pounds did the tusk weigh?

Brain Builders

18. **Processes &Practices** **Reason** Which is greater: 3^5 or 5^3? Explain your reasoning.

19. **Building on the Essential Question** What does it mean for a product of factors to be expressed using exponents?

Name ....

Homewwork Helper

Need help? connectED.mcgraw-hill.com

Write 6 × 6 × 6 using an exponent.

The base is 6. Since 6 is used as a factor three times, the exponent is 3.

So, $6 \times 6 \times 6 = 6^3$.

Practice

Write each product using an exponent.

1. $10 \times 10 \times 10 =$ _____

2. $12 \times 12 =$ _____

Write each power as a product of the same factor. Then find the value.

3. $3^7 =$ _____

4. $10^6 =$ _____

Write the prime factorization of each number using exponents.

5. $20 =$ _____

6. $50 =$ _____

Problem Solving

7. **Processes & Practices** **8** **Look for a Pattern** The Newfoundland is a large breed of dog. It weighs about 10 × 10 pounds. Write 10 × 10 using an exponent. Then find the value of the power. How many pounds does the Newfoundland weigh?

_____

Brain Builders

8. The area of San Bernardino County, California, the largest county in the United States, is about 3^9 square miles. Write this as an expression. What is the area of San Bernardino County?

_____

Is $3^3 + 3^3 + 3^3$ the same as 3^9? Explain why or why not.

_____

Vocabulary Check

Fill in each blank with the correct term or number to complete each sentence.

9. Numbers expressed using exponents are called _____.

10. The exponent indicates how many times the _____ is used as a factor.

11. **Test Practice** A 100-pound person on Earth would weigh about 4 × 4 × 4 × 4 pounds on Jupiter. Evaluate the expression to determine how much a 100-pound person would weigh on Jupiter. How much would a 200-pound person weigh?

Ⓐ 16 pounds ; 32 pounds

Ⓒ 256 pounds ; 512 pounds

Ⓑ 64 pounds ; 32 pounds

Ⓓ 1,024 pounds ; 2,048 pounds

Name

ESSENTIAL QUESTION
What strategies can be used to multiply whole numbers?

Powers of 10 include numbers like 10, 100, and 1,000, because they can be written as 10^1, 10^2, and 10^3.

 ## Math in My World [Watch ▶] [Tutor 💬]

Example 1

Pet Paws is ordering more goldfish to sell at their store. Each goldfish costs $2. Use the table to determine the cost of 10, 100, and 1,000 goldfish. Describe the pattern in the number of zeros when multiplying the cost, $2, by powers of ten.

Cost of One Goldfish ($)	Power of Ten	Product	Number of Zeros in Product
2	× 1	2	0
2	× 10		
2	× 100		
2	× 1,000		

The number of zeros in the product increases when the power of ten increases. Each successive power of ten adds

_____ zero to the product.

How many zeros are in the product of 7 and 100? _____

How many zeros are in the product of 21 and 10? _____

How many zeros are in the product of 12 and 1,000? _____

Example 2

Find 13×10^2 mentally.

 Write 10^2 without exponents.

$13 \times 10^2 =$

$13 \times$ _____

 Count the number of zeros in the power of 10.

100 ← [_____ **zeros**]

Compare the number of zeros to the exponent of 10^2.

They are the _____.

 Write the zeros to the right of 13.

1, 3 ____

So, the product is

_____.

Example 3

Find $40 \times 7{,}000$ mentally.

 Write the basic multiplication fact.

$4 \times$ _____ $= 28$

Count the number of zeros in each factor.

$40 \times 7{,}000$

[**1 zero and** ____ **zeros**]

There are a total of

_____ zeros.

Write the zeros to the right of the product from step 1.

28__, ____

So, the product is

_____.

Guided Practice

Find each product mentally.

1. 8×10^2

$10^2 =$ _____

total number of zeros = _____

So, $8 \times 10^2 =$ _____.

2. $14 \times 2{,}000$

basic multiplication fact:

_____ $\times$ _____ $= 28$

total number of zeros = _____

So, $14 \times 2{,}000 =$ _____.

Explain how you could find the product of 29 and 10^3 mentally.

Independent Practice

Find each product mentally.

3. $13 \times 1{,}000 =$ _____

4. $37 \times 10^2 =$ _____

5. $9{,}000 \times 3 =$ _____

6. $8 \times 10^3 =$ _____

7. $21 \times 10^1 =$ _____

8. $9 \times 50 =$ _____

Algebra Find the missing number.

9. $\blacksquare \times 10^4 = 70{,}000$

$\blacksquare =$ _____

10. $300 \times \blacksquare = 120{,}000$

$\blacksquare =$ _____

11. $100 \times \blacksquare = 900$

$\blacksquare =$ _____

12. $\blacksquare \times 10^2 = 4{,}400$

$\blacksquare =$ _____

Problem Solving

13. Each box contains 10^2 pencils. The school store has 15 boxes of pencils. How many pencils does the school store have?

_____

14. **Processes &Practices** 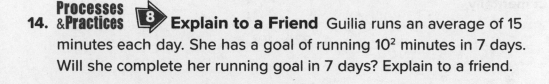 **8** **Explain to a Friend** Guilia runs an average of 15 minutes each day. She has a goal of running 10^2 minutes in 7 days. Will she complete her running goal in 7 days? Explain to a friend.

_____

_____

15. A group of friends bought 7 tickets to a dog show for $30 each. How much did they spend on the tickets?

_____

Processes &Practices **2** **Use Number Sense** Find each missing exponent.

16. $40 \times 10^\blacksquare = 4,000$ **17.** $32 \times 10^\blacksquare = 32,000$ **18.** $80,000 = 10^\blacksquare \times 8$

$\blacksquare = $ _____ $\blacksquare = $ _____ $\blacksquare = $ _____

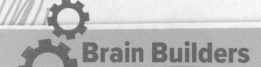 **Brain Builders**

19. Explain how to determine the number of zeros in the product of $10,000 \times 1,000$.

_____

_____

20. **?** **Building on the Essential Question** How can patterns be used to determine products of a number and a power of 10?

_____

_____

_____

Name _____

MY Homework

Lesson 4

Multiplication Patterns

Homework Helper

Need help? connectED.mcgraw-hill.com

A truck is loaded with 10^2 boxes of skateboards. Each box weighs 36 pounds. What is the total weight of the boxes?

Find 36×10^2 mentally.

1 10^2 without exponents is equal to 100.

2 There are 2 zeros in 100.

3 After placing the zeros to the right of 36, the product is 3,600.

So, the total weight of the boxes is 3,600 pounds.

Practice

Find each product mentally.

1. $70 \times 500 = $ _____

2. $320 \times 10^2 = $ _____

3. $56 \times 10^3 = $ _____

4. $10^2 \times 72 = $ _____

5. $80 \times 3,000 = $ _____

6. $10^3 \times 31 = $ _____

Problem Solving

7. To protect themselves from extreme hot or cold temperatures, American Alligators dig burrows in the mud. Suppose there are 20 alligators, each with 50 feet of burrows. What is the total length of all the burrows?

_____

8. Paulita reads an average of 20 pages each day. She has 6 days to read 10^2 pages. Will she finish her reading in 6 days? Explain.

_____

_____

Brain Builders

9. **Processes &Practices** **Make Sense of Problems** Explain how using basic facts can help you find $10 \times 20 \times 30 \times 40$ mentally.

_____

_____

Vocabulary Check

Fill in the blank with the correct term or number to complete the sentence.

10. Numbers like 10, 100, 1,000, and so on are called

_____ .

11. **Test Practice** A music store sold 10^3 CDs and 10^2 CD players. If each CD costs \$12 and each CD player costs \$35, what was the store's total earnings?

Ⓐ \$15,500

Ⓒ \$36,200

Ⓑ \$24,500

Ⓓ \$47,000

Lesson 5
Problem-Solving Investigation
STRATEGY: Make a Table

ESSENTIAL QUESTION
What strategies can be used to multiply whole numbers?

Learn the Strategy

 Watch   Tutor

In a recent year, about 63 out of every 100 households in the United States owned at least one pet. About how many households out of ten thousand owned at least one pet?

1 Understand

What facts do you know?

Sixty-three out of every _____ households owned at least one pet.

What do you need to find?

how many households out of ten thousand owned at least one pet

2 Plan

I will make a _____ to solve the problem.

3 Solve

Total Number of Households	Number of Households Owning at least one Pet
100	63

×10 ×10
×10 ×10

So, about _____ out of ten thousand households owned at least one pet.

4 Check

Is my answer reasonable? Explain.

Use mental math to multiply.   63 × 100 = _____

Nestor is saving money to buy a new camping tent. Each week he doubles the amount he saved the previous week. If he saves $1 the first week, how much money will Nestor save in 7 weeks?

 Understand

What facts do you know?

_____

_____

What do you need to find?

_____

 **Plan**

_____

_____

Solve

Check

Is my answer reasonable? Explain.

_____

_____

Apply the Strategy

Solve each problem by making a table.

1. Betsy is saving to buy a bird cage. She saves $1 the first week, $3 the second week, $9 the third week, and so on. How much money will she save in 5 weeks?

_____

2. Kendall is planning to buy a laptop for $1,200. Each month she doubles the amount she saved the previous month. If she saves $20 the first month, in how many months will Kendall have enough money to buy the laptop?

_____

Brain Builders

3. Mrs. Piant's yearly salary is $42,000 and increases $2,000 per year. Mr. Piant's yearly salary is $37,000 and increases $3,000 per year. In how many years will Mr. and Mrs. Piant make the same salary? Explain how you used a table to solve.

_____

_____

4. **Processes &Practices**  **Look for a Pattern** Complete the table that shows the prime factorizations of powers of 10. Use the pattern in the table to mentally find the prime factorization of 10^9. Write using exponents.

Power of 10	Prime Factorization
10	2×5
100 or 10^2	$2^2 \times 5^2$
1,000 or 10^3	
10,000 or 10^4	

Prime factorization of $10^9 =$ _____ .

Review the Strategies

Use any strategy to solve each problem.
- Make a table.
- Use the four-step plan.

A card shop recorded how many packs of trading cards it sold each day. Use the table to solve Exercises 5–7.

Trading Cards Sold			
Day	Week 1	Week 2	Week 3
Mon.	28	48	25
Tue.	32	43	37
Wed.	38	45	42
Thur.	44	41	35
Fri.	36	39	41

5. In which week did they sell the most packs of cards?

_____

6. In which week did they sell the least amount?

_____

7. How many more packs did they sell in Week 2 than in Week 3?

_____

Solve.

8. A putt-putt course offers a deal that when you purchase 10 rounds of golf, you get 1 round for free. If you played a total of 35 rounds, how many rounds did you purchase?

_____

9. **Processes &Practices** **1** **Plan Your Solution** Saketa is saving money to buy a new ferret cage. In the first week, she saved $24. Each week after the first, she saves $6. How much money will Saketa have saved in six weeks?

_____

Name

MY Homework

Homework Helper eHelp

Need help? connectED.mcgraw-hill.com

Every 10 minutes, 160 passengers can ride the Black Stallion roller coaster. How many people can ride the Black Stallion in 60 minutes?

1 Understand

What facts do you know?

There are 160 passengers that ride every 10 minutes.

What do you need to find?

the number of people that can ride the coaster in 60 minutes

2 Plan

I can make a table to solve the problem.

3 Solve

Minutes	10	20	30	40	50	60
Passengers	160	320	480	640	800	960

+160 +160 +160 +160 +160

So, 960 passengers can ride the Black Stallion in 60 minutes.

4 Check

Is my answer reasonable? Explain.

Estimate. Round 160 to the nearest hundred.

200 + 200 + 200 + 200 + 200 + 200 = 1,200

![Real World](globe icon) **Problem Solving**

Solve each problem by making a table.

Processes
1. &Practices **8** **Look for a Pattern** In a recent year, one United States dollar was equal to about 82 Japanese yen. How many Japanese yen are equal to $100? $1,000? $10,000?

_____

2. A local restaurant offers a deal if you purchase 3 medium pizzas, you get 2 side dishes for free. If you get a total of 8 side dishes, how many pizzas did you buy?

_____

3. A recipe for potato salad calls for one teaspoon of vinegar for every 2 teaspoons of mayonnaise. How many teaspoons of vinegar are needed for 16 teaspoons of mayonnaise?

_____

Brain Builders

4. A package of 4 mechanical pencils comes with 2 free erasers. If you get a total of 12 free erasers, how many packages of pencils did you buy? If you get 18 free erasers, how many packages of pencils did you buy?

_____

5. Veronica is saving money to buy a saddle for her horse that costs $175. She plans to save $10 the first month and then increase the amount she saves by $5 each month after the first month. How many months will it take her to save $175? Propose a way she can save the same amount, but in two fewer months.

_____

Check My Progress

Vocabulary Check

Draw a line to match each word to its definition.

1. Numbers such as 1, 2, 4, 5, 10, and 20.

2. A number raised to the second power.

3. A number raised to the third power.

4. A way of expressing a composite number as a product of its prime factors.

5. Numbers like 10, 100, 1,000 and so on.

- cubed

- prime factorization

- factors

- powers of ten

- squared

Concept Check

Find the prime factorization of each number using exponents.

6. 36 = _____

7. 45 = _____

Write each product using an exponent.

8. 13 × 13 × 13 × 13 = _____

9. 7 × 7 × 7 = _____

Write each power as a product of the same factor. Then find the value.

10. $10^4 = $ _____

11. $14^1 = $ _____

Find each product mentally.

12. $23 \times 100 = $ _____

13. $150 \times 400 = $ _____

Problem Solving

14. The amount of space in the dog carrier shown can be found by multiplying its width, length, and height. Write the amount of space, in cubic units, using an exponent. Then evaluate.

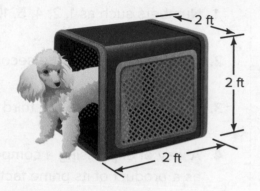

2 ft

2 ft

2 ft

_____

Brain Builders

15. There are 10^2 fifth-grade students attending a class field trip to the city's art museum. If each admission ticket costs $6, what is the total cost of admission for the 10^2 students? Explain how you know.

_____

_____

16. Test Practice Dakota read 130 pages over the weekend. Dajuan read 165 pages over the weekend. Which prime number do the prime factorizations of 130 and 165 have in common?

Ⓐ 2

Ⓒ 5

Ⓑ 3

Ⓓ 7

Name _____

Lesson 6
Hands On
Use Partial Products and the Distributive Property

Draw It [Tools]

Darnel and his four friends went to an ice skating rink and bought sandwiches. They divided the total cost and found that each person needs to pay $17. What was the total cost for ice skating and sandwiches?

Find 5 × 17 using an area model.

1 Label the model to find the partial products.

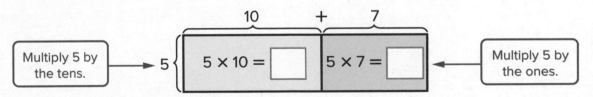

	10	+	7
Multiply 5 by the tens. →	5 { 5 × 10 = ☐		5 × 7 = ☐	← Multiply 5 by the ones.

2 Add the partial products.

_____ + _____ = _____

So, the total cost for ice skating and sandwiches was _____.

When you use partial products, you are using a property. A **property** is a rule in mathematics that can be applied to all numbers. The property that you applied above is called the **Distributive Property.** You will learn more about this property in the next lesson.

Find 7 × 56 using an area model.

1. Label the model to find partial products.

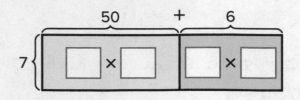

2. Multiply. Then add.

   7 × 56 = (7 × 50) + (7 × _____)

   = _____ + _____

   = _____

So, 7 × 56 = _____.

Talk About It

1. **Processes & Practices** 3 **Draw a Conclusion** How do area models show partial products?

   _____

   _____

   _____

2. In finding 7 × 56 above, why does the area model separate 56 into 50 and 6?

   _____

   _____

3. Find 6 × 36 using partial products. Label the model to help you solve. Show your work.

   _____

   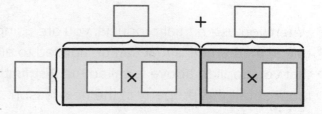

   _____

Practice It

Multiply using the area model. Label each model.

4. 4 × 16 = _____

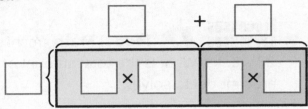

5. 6 × 81 = _____

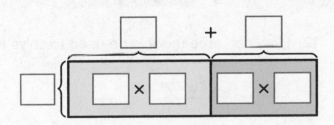

6. 7 × 29 = _____

7. 5 × 39 = _____

8. Ronnie swims 4 laps each day at the pool. How many laps does he swim in 28 days? Use an area model to solve.

9. **Processes & Practices** **4** **Model Math** Yoshi saved $5 each week for 23 weeks. What is the total amount of money he saved? Use an area model to solve.

10. Thirty-eight fish are in each aquarium at a fish store. How many fish are there in five aquariums? Use an area model to solve, then write the equation from your model

11. **Processes & Practices** **3** **Find the Error** Omar was using an area model to find 4 × 61. Find his mistake and correct it.

$$(4 \times 60) + (4 \times 10) = 240 + 40 = 280$$

Write About It

12. How can area models be used to solve multiplication problems?

 **MY Homework**

Homework Helper

Need help? ⟿ connectED.mcgraw-hill.com

Consuelo and her two siblings decided to buy a trampoline.
They divided the total cost and found that each person needs
to pay $48. What was the total cost for the trampoline?

Find 3 × 48 using an area model.

1 The model is labeled to show
the partial products.

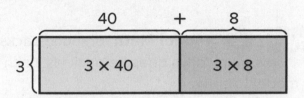

2 Multiply. Then add.

$$3 \times 48 = (3 \times 40) + (3 \times 8)$$
$$= 120 + 24$$
$$= 144$$

So, the total cost of the trampoline was $144.

Practice

1. Multiply 2 × 27 using the area model. Label the model.

2 × 27 = _____

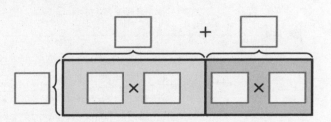

2. Multiply 9 × 43 using the area model. Label the model.

9 × 43 = _____

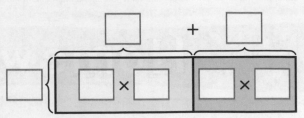

 Problem Solving

3. A pet store has eighteen hamsters in each cage. How many hamsters are there in six cages? Use an area model to solve.

_____

4. **Processes &Practices** ▶ **4** **Model Math** Brandon packs his lunch 4 days each week. How many days does he pack his lunch for 36 weeks? Use an area model to solve.

_____

Brain Builders

5. Ricarda earns $25 allowance each week. How can she mentally compute her total earnings after 7 weeks?

_____

_____

6. Write an expression that uses partial products to multiply 8 × 64. Can your expression be written another way? Explain.

_____

_____

_____

Name _____

ESSENTIAL QUESTION
What strategies can be used to multiply whole numbers?

The **Distributive Property** allows you to multiply a sum by a number. To do so, multiply each addend by the number. Then add.

$$3 \times (5 + 2) = (3 \times 5) + (3 \times 2)$$

Math in My World

Watch | Tools | Tutor

Example 1

For a field trip, 42 students each paid $3 for transportation. Use mental math and the Distributive Property to find how much money was collected altogether.

Find 3 × 42.

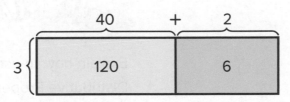

	40	+	2
3 {	120		6

1 Write 42 as 40 + 2.

$3 \times 42 = 3 \times ($ _____ $+$ _____ $)$

2 Apply the Distributive Property.

$= (3 \times$ _____ $) + (3 \times$ _____ $)$

3 Multiply.

$=$ _____ $+$ _____

4 Add.

$=$ _____

So, $ _____ was collected for the field trip.

Example 2

Find 7 × 26 mentally using the Distributive Property.
Show the steps that you used.

 Write 26 as 20 + 6.

$$7 \times 26 = 7 \times (\underline{\hspace{1cm}} + \underline{\hspace{1cm}})$$

 Apply the Distributive Property.

$$= (7 \times \underline{\hspace{1cm}}) + (7 \times \underline{\hspace{1cm}})$$

 Multiply.

$$= \underline{\hspace{1cm}} + \underline{\hspace{1cm}}$$

 Add.

$$= \underline{\hspace{1cm}}$$

So, 7 × 26 = _____.

Guided Practice

Find each product mentally using the Distributive Property.
Show the steps that you used.

1. 5 × 18

$$5 \times 18 = 5 \times (\underline{\hspace{1cm}} + 8)$$
$$= (5 \times \underline{\hspace{1cm}}) + (5 \times \underline{\hspace{1cm}})$$
$$= \underline{\hspace{1cm}} + \underline{\hspace{1cm}}$$
$$= \underline{\hspace{1cm}}$$

So, 5 × 18 = _____.

Talk MATH

Explain how to use the Distributive Property to find a product mentally.

2. 2 × 21

$$2 \times 21 = 2 \times (\underline{\hspace{1cm}} + \underline{\hspace{1cm}})$$
$$= (2 \times \underline{\hspace{1cm}}) + (2 \times \underline{\hspace{1cm}})$$
$$= \underline{\hspace{1cm}} + \underline{\hspace{1cm}}$$
$$= \underline{\hspace{1cm}}$$

So, 2 × 21 = _____.

Independent Practice

Find each product mentally using the Distributive Property. Show the steps that you used.

3. $6 \times 13 = $ _____

4. $3 \times 52 = $ _____

5. $5 \times 26 = $ _____

6. $4 \times 69 = $ _____

7. $2 \times 49 = $ _____

8. $7 \times 23 = $ _____

9. $26 \times 6 = $ _____

10. $55 \times 4 = $ _____

Problem Solving

11. A horse is 17 *hands* tall. If a *hand* equals 4 inches, how tall is the horse in inches? Use the Distributive Property to solve.

_____

12. Mr. Collins is buying 5 train tickets for $36 each. What is the total cost of the tickets? Use the Distributive Property to solve.

_____

13. Melanie runs 23 miles each week. How many miles does she run in 9 weeks? Use the Distributive Property to solve.

_____

Brain Builders

14. **Processes &Practices** **3** **Find the Error** Dylan is using the Distributive Property to simplify $6 \times (9 + 4)$. Find his mistake and correct it. Explain to a friend.

$$6 \times (9 + 4) = 54 + 4$$
$$= 58$$

_____

_____

_____

15. **Building on the Essential Question** How can the Distributive Property be used to multiply numbers? Explain.

_____

_____

Name

 **MY Homework**

Homework Helper

Need help? connectED.mcgraw-hill.com

At a petting zoo, 29 students each paid **$2** to pet the animals. Use mental math and the Distributive Property to find how much money was paid altogether.

1 Write 29 as 20 + 9. ⟶ 2 × 29 = 2 × (20 + 9)

2 Apply the Distributive Property. ⟶ = (2 × 20) + (2 × 9)

3 Multiply. ⟶ = 40 + 18

4 Add. ⟶ = 58

So, the students spent $58 at the petting zoo.

Practice

Find each product mentally using the Distributive Property. Show the steps that you used.

1. 4 × 48 = _____

2. 3 × 67 = _____

3. 6 × 18 = _____

4. 8 × 74 = _____

Problem Solving

5. A bookshelf in Deirdre's living room has 4 shelves. There are 12 books on each shelf. How many books are on the bookshelf altogether?

_____

6. Jorge is collecting baseball cards. He has 29 stacks of cards with 4 in each stack. How many cards does he have altogether?

_____

Brain Builders

7. **Processes &Practices** **3** **Justify Conclusions** The Distributive Property also combines subtraction and multiplication. For example, $3 \times (5 - 2) = (3 \times 5) - (3 \times 2)$. Explain how you could use the Distributive Property and mental math to find 5×198.

_____

_____

Vocabulary Check

Fill in each blank with the correct term or number to complete the sentence.

8. The Distributive Property combines _____ and

_____ to make multiplying whole numbers simpler.

9. **Test Practice** Susana collected 5 cents at the recycling plant for each of her 78 cans. How much money did she collect altogether? Explain how you know one of the other answer choices is incorrect.

    Ⓐ $0.39   Ⓒ $3.90

    Ⓑ $3.50   Ⓓ $39.00

_____

_____

_____

Name ....

ESSENTIAL QUESTION
What strategies can be used to multiply whole numbers?

When a problem asks for *about* how many, you can use estimation, rounding, and/or compatible numbers. **Compatible numbers** are numbers in a problem that are easy to compute mentally.

 Math in My World Watch Tutor

Example 1

A pet store has 12 gecko lizards for sale. Each gecko lizard costs $92. About how much money would the store make if it sells all 12 gecko lizards?

Estimate the product of 92 and 12.

One Way Round one factor.

THINK It is easier to compute 92 × 10 than 90 × 12.

$$92 \longrightarrow 9\ 2$$
$$\times 12 \longrightarrow \times 1\ \boxed{\ }$$ Round 12 to the nearest ten.

$$\boxed{\ }\ \boxed{\ }\ \boxed{\ }$$ Find 92 × 10 mentally.

By rounding one factor, the estimate is _____.

Another Way Round both factors.

$$92 \longrightarrow 9\ \boxed{\ }$$ Round 92 to the nearest ten.
$$\times 12 \longrightarrow \times 1\ \boxed{\ }$$ Round 12 to the nearest ten.

$$\boxed{\ }\ \boxed{\ }\ \boxed{\ }$$ Find 90 × 10 mentally.

By rounding both factors, the estimate is _____.

Copyright © McGraw-Hill Education G.K. & Vikki Hart/Photodisc/Getty Images

Another Way   Use compatible numbers.

$$92 \longrightarrow \boxed{}\boxed{}\boxed{}$$
$$\underline{\times\ 12} \longrightarrow \underline{\times\ 1\ \boxed{}}$$
$$\boxed{},\boxed{}\boxed{}\boxed{}$$

Use numbers that are easy to multiply mentally such as 100 and 10.

Find 100 × 10 mentally.

By using compatible numbers, the estimate is _____.

Use a calculator to multiply 92 × 12. What do you get? _____

Circle whether the estimates were all less than or greater than the actual product.

less than   greater than

Guided Practice

Estimate by rounding or using compatible numbers. Show how you estimated.

1. Round both factors.

$$32 \longrightarrow \quad 3\ \boxed{}$$
$$\underline{\times\ 18} \longrightarrow \underline{\times\ 2\ \boxed{}}$$
$$\boxed{}\boxed{}\boxed{}$$

The product is about _____.

Talk MATH

Show two different ways you could estimate 312 × 18.

2. Use compatible numbers.

$$98 \longrightarrow \boxed{}\boxed{}\boxed{}$$
$$\underline{\times\ 83} \longrightarrow \underline{\times\ \boxed{}\boxed{}}$$
$$\boxed{},\boxed{}\boxed{}\boxed{}$$

The product is about _____.

126   Chapter 2 Multiply Whole Numbers

Name ....

Estimate by rounding. Show how you estimated.

3.   218
        ×   6

4.   68
        ×   7

5.   131
        × 29

6. 61 × 68 ≈ _____

7. 79 × 56 ≈ _____

8. 392 × 46 ≈ _____

Estimate by using compatible numbers. Show how you estimated.

9.   106
        ×   52

10.   33
         ×   6

11.   127
         ×   8

12. 33 × 84 ≈ _____

13. 450 × 21 ≈ _____

14. 729 × 42 ≈ _____

Problem Solving

15. The table shows the number of pounds of apples that were harvested each day. Estimate how many pounds of apples were harvested altogether by rounding. Show how you estimated.

Day	Pounds of Apples
1	514
2	487
3	349
4	421
5	392

_____

_____

16. In one week, a campground rented 18 cabins at $225 each. About how much did they collect in rent altogether? Show how you estimated.

_____

17. The average weight for a guinea pig is 2 pounds. The average weight for a pot belly pig is 45 times greater than the guinea pig. About how much does an average pot belly pig weigh? Show how you estimated.

_____

Brain Builders

18. **Processes &Practices** ➡ **2** **Use Number Sense** Use the digits 1, 3, 4, and 7 to create two whole numbers whose product is estimated to be about 600. Explain your method.

_____

_____

19. **Building on the Essential Question** When is the estimation of products a useful tool? Provide an example.

_____

_____

_____

Name ....

MY Homework

Homework Helper

Need help? ➚ **connectED.mcgraw-hill.com**

Mountain View Elementary is sending **21** boxes of magazines to a school in Uruguay. There are **154** magazines in each box. About how many magazines are they sending?

South America

Uruguay

Estimate the product of 21 and 154.

One Way Round each factor to the nearest ten.

$$
\begin{array}{r}
154 \longrightarrow 150 \\
\times\ \ 21 \longrightarrow \times\ \ 20 \\
\hline
3{,}000
\end{array}
$$

Round 154 to the nearest ten.

Round 21 to the nearest ten.

Find 150 × 20 mentally.

By rounding both factors to the nearest ten, the estimate is about

3,000 magazines.

Another Way Use compatible numbers.

$$
\begin{array}{r}
154 \longrightarrow 200 \\
\times\ \ 21 \longrightarrow \times\ \ 20 \\
\hline
4{,}000
\end{array}
$$

Use numbers that are easy to multiply mentally such as 200 and 20.

Find 200 × 20 mentally.

By using compatible numbers, the estimate is about

4,000 magazines.

Practice

Estimate by rounding or using compatible numbers. Show how you estimated.

1. $\begin{array}{r} 4 \\ \times\ 24 \\ \hline \end{array}$

2. $\begin{array}{r} 76 \\ \times\ 78 \\ \hline \end{array}$

3. $\begin{array}{r} 508 \\ \times\ 27 \\ \hline \end{array}$

Problem Solving

4. For a school assembly, students sit in chairs that are arranged in 53 rows. There are 12 chairs in each row. About how many students can be seated? Show how you estimated.

_____

5. Klara bought a dozen bags of bird food for $27. Use compatible numbers to find the approximate cost of six dozen bags of bird food. Show how you estimated.

_____

Brain Builders

6. **Processes &Practices** **Find the Error** Rico is estimating 139 × 18. Find his mistake and correct it. Explain how you know.

$$100 \times 10 = 1,000$$

_____

_____

Vocabulary Check

Fill in the blank with the correct term or number to complete the sentence.

7. Compatible numbers are numbers in a problem that are easy to compute _____.

8. **Test Practice** On a cross-country vacation, Maria filled her 14-gallon gas tank eleven times. Which is the best estimate of how many gallons of gas she put in the tank altogether?

Ⓐ 75 gallons   Ⓒ 200 gallons

Ⓑ 150 gallons   Ⓓ 225 gallons

Lesson 9
Multiply by One-Digit Numbers

ESSENTIAL QUESTION
What strategies can be used to multiply whole numbers?

Math in My World

Watch | Tools | Tutor

Example 1

Grace and her three friends each paid $38 for an admission ticket to an amusement park. The total paid can be found by multiplying 4 and 38.

Find 38 × 4.

1 **Multiply the ones.**

8 ones × 4 = 32 ones

Regroup 32 ones as
3 tens and 2 ones.

2 **Multiply the tens.**

3 tens × 4 = 12 tens

Add any new tens.

12 tens + 3 tens = 15 tens

$$\begin{array}{r} \square \\ 3\ 8 \\ \times\quad 4 \\ \hline \square\ \square\ \square \end{array}$$

So, the total amount paid for admission to

an amusement park is _____ .

Check You can use an area model to check your answer.

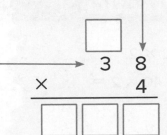

	30	+	8
4	120		32

Example 2

Find 317 × 5.

Estimate 300 × 5 = _____

 Multiply the ones.

7 ones × 5 = 35 ones

Regroup 35 ones as 3 tens and 5 ones.

2 Multiply the tens.

1 ten × 5 = 5 tens

Add any new tens.

5 tens + 3 tens = 8 tens

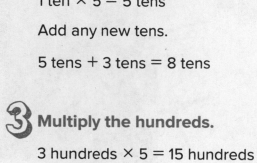

```
      3   1   7
  ×           5
 ┌──┐ ┌──┐┌──┐┌──┐
 │  │ │  ││  ││  │
 └──┘ └──┘└──┘└──┘
    ,
```

3 Multiply the hundreds.

3 hundreds × 5 = 15 hundreds

So, 317 × 5 = _____ .

Check Compare to the estimate. _____ ≈ _____

Guided Practice

1. Multiply 42 × 2.

   Estimate 40 × 2 = _____

```
      4 2
  ×     2
  ┌──┐┌──┐
  │  ││  │
  └──┘└──┘
```

So, 42 × 2 = _____ .

Check

_____ ≈ _____

Describe each step for finding 416 × 3.

Independent Practice

Estimate. Then multiply. Use your estimate to check your answer.

2. 21
    × 3

3. 32
    × 6

4. 52
    × 9

5. 401
    ×   7

6. 712
    ×   3

7. 413
    ×   9

8. 31 × 5 = _____

9. 208 × 3 = _____

10. 47 × 6 = _____

11. 211 × 7 = _____

12. 182 × 6 = _____

13. 806 × 7 = _____

Problem Solving

14. One 747-airplane can carry 420 passengers. How many total passengers can three planes carry?

_____

15. **Processes &Practices**  **⑤ Use Math Tools** In an auditorium, there are 9 rows of seats with 18 seats in each row. There are also 6 rows of seats with 24 seats in each row. How many seats are there in the auditorium? Estimate first. Then check for reasonableness.

_____

16. Malia bought 14 cans of cat food. Each can weighs 8 ounces. How many total ounces of cat food did she buy?

_____

Brain Builders

17. **Processes &Practices** **② Use Number Sense** Catalina multiplied 842 and 3 and got 3,326. How can she check to see if her answer is reasonable? Write a number sentence to support your answer.

_____

_____

_____

18. **? Building on the Essential Question** What is the procedure for multiplying by a one-digit number?

_____

_____

Name ....

MY Homework

Homework Helper

Need help? **connectED.mcgraw-hill.com**

The world's largest cactus is 5 times as tall as the cactus shown. How tall is the world's largest cactus?

Find 15 × 5.

Estimate 20 × 5 = 100

1 **Multiply the ones.**

5 ones × 5 = 25 ones

Regroup 25 ones as 2 tens and 5 ones.

2 **Multiply the tens.**

1 ten × 5 = 5 tens

Add any new tens.

5 tens + 2 tens = 7 tens

$$
\begin{array}{r}
2 \\
15 \\
\times\ 5 \\
\hline
75
\end{array}
$$

So, the world's largest cactus is 75 feet tall.

Check Compare to the estimate. 75 ≈ 100

15 ft

Practice

Estimate. Then multiply. Use your estimate to check your answer.

1.   18
     × 8

2.   72
     × 4

3.   341
     × 4

Processes & Practices ➔ ③ **Check for Reasonableness** Malcolm ran the 440-yard dash and the 220-yard dash at a track meet. There are 3 feet in one yard. How many total feet did Malcolm run? Estimate first. Then check for reasonableness.

5. Each student in Mrs. Henderson's science class brought in 3 books for the book donation. If there are 25 students in the class, how many total books did they collect?

6. Karen and Anthony are setting up rows for the piano recital. They set up 24 rows with 6 chairs in each row. How many total people will the rows seat?

Brain Builders

7. Veronica brought her turtle out of its aquarium for 15 minutes every night for 7 days. How many total minutes did she bring her turtle out of its aquarium? Write another scenario that uses the same numbers.

8. **Test Practice** A restaurant has 36 tables. If each table can sit five people, how many people can be seated at the restaurant? How many people sit at each table if there are 288 people seated?

   Ⓐ 216 people; 6 people

   Ⓑ 180 people; 8 people

   Ⓒ 150 people; 6 people

   Ⓓ 41 people; 8 people

Name _____

Math in My World

 Watch ▶ | Tutor 💬

Example 1

Domestic cats can run up to 44 feet per second on land. At this rate, how many feet could a cat run in 12 seconds?

Find 44 × 12.

Estimate 44 × 10 = _____

1 **Multiply the ones.**

44 × 2 = 88

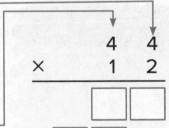

2 **Multiply the tens.**

44 × 10 = 440

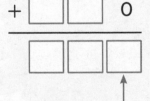

$$+ \quad \boxed{} \ \boxed{} \ 0$$

Helpful Hint

By estimating first, you can determine if your answer is reasonable.

3 **Add.** _____

88 + 440 = 528

So, a domestic cat can run _____ feet in 12 seconds.

Check Compare to the estimate. _____ ≈ _____

Example 2

Find 165 × 31.

Estimate 200 × 30 = _____

 Multiply the ones.

165 × 1 = 165

 Multiply the tens.

165 × 30 = 4,950

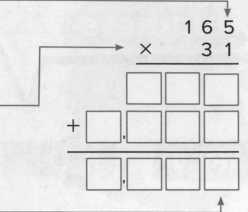

$$
\begin{array}{r}
1\ 6\ 5 \\
\times \quad 3\ 1 \\
\hline
\square\ \square\ \square \\
+\ \square,\square\ \square\ \square \\
\hline
\square,\square\ \square\ \square \\
\end{array}
$$

 Add.

165 + 4,950 = 5,115

So, 165 × 31 = _____ .

Check Compare to the estimate. _____ ≈ _____

Guided Practice

1. Multiply 32 × 13.

Estimate 30 × 10 = _____

$$
\begin{array}{r}
3\ 2 \\
\times \quad 1\ 3 \\
\hline
\square\ \square \\
+\ \square\ \square\ 0 \\
\hline
\square\ \square\ \square \\
\end{array}
$$

So, 32 × 13 = _____ .

Check for Reasonableness

_____ ≈ _____

Describe how addition is used when you multiply by two-digit numbers.

Name _____

Independent Practice

Estimate. Then multiply. Use your estimate to check your answer.

2. 102
    × 12

3. 102
    × 56

4. 24
    × 21

5. 39
    × 34

6. 13
    × 54

7. 51
    × 82

8. 21 × 42 = _____

9. 69 × 14 = _____

10. 83 × 367 = _____

11. 534 × 67 = _____

12. 141 × 25 = _____

13. 229 × 31 = _____

14. Marshall's mother buys 2 boxes of granola bars each week. Each box contains 8 granola bars. If she continues buying 2 boxes each week, how many granola bars will she buy in a year (1 year = 52 weeks)?

_____

15. **Processes &Practices** **Use Math Tools** A delivery truck travels 278 miles each day. How far does it travel in 25 days?

_____

Brain Builders

16. A farmer buys 1,000 pounds of hay, which he thinks will last 31 days. If his cow eats about 25 pounds of hay per day, for how many days more than 31 will the farmer's hay feed the cow? Explain.

_____

_____

17. **Processes &Practices** **2** **Use Number Sense** Find 235 × 124. Use the same strategy that you used for multiplying by a two-digit number except include multiplying by the hundreds place. Show your work.

_____

_____

18. **Building on the Essential Question** How do you multiply by two-digit numbers?

_____

_____

Name _____

MY Homework

Homework Helper

Need help? connectED.mcgraw-hill.com

Alicia lives in Nashville, Tennessee. Last year her family drove to Atlanta, Georgia, each month to visit her grandmother. Find the total distance they drove to visit her grandmother for the year.

Destination City From Nashville	Round-Trip Distance (mi)
Atlanta	498
Raleigh	1,080

Find 498 × 12.

Estimate 500 × 10 = 5,000

1 Multiply the ones.

498 × 2 = 996

2 Multiply the tens.

498 × 10 = 4,980

```
    498
  ×  12
  ─────
    996
+ 4,980
  ─────
  5,976
```

3 Add.

996 + 4,980 = 5,976

So, they drove a total of 5,976 miles for the year.

Check Compare to the estimate. 5,976 ≈ 5,000

Practice

Estimate. Then multiply. Use your estimate to check your answer.

1. 19
  × 15

2. 43
  × 65

3. 470 × 56 = _____

4. Ms. Jenkins was arranging chairs for a school awards assembly. Each row contained 15 chairs. If there were 21 rows, how many chairs had to be arranged?

_____

5. Leon earns $14 an hour. How much does he earn in 4 weeks if he works 12 hours each week?

_____

Brain Builders

6. **Processes &Practices** **5** **Use Math Tools** Without actually calculating, how much greater is the product of 98 × 50 than the product of 97 × 50? Explain.

_____

_____

7. The table shows Katrina's prices for dog walking. If she walks 5 medium-sized dogs and 8 large-sized dogs for 12 weeks, how much will she earn?

Dog Type	Cost Per Week ($)
Small	10
Medium	12
Large	14

_____

8. **Test Practice** Each day there are 12 tours at the glass factory. Twenty-eight people can go on a tour. How many people can tour the glass factory each day?

Ⓐ 236 people   Ⓒ 336 people

Ⓑ 280 people   Ⓓ 436 people

Processes &Practices 6

Multiply.

1. 17
    × 6

2. 24
    × 7

3. 31
    × 3

4. 68
    × 2

5. 41
    × 8

6. 92
    × 5

7. 19
    × 5

8. 67
    × 7

9. 32
    × 4

10. 90
    × 6

11. 83
    × 2

12. 62
    × 5

13. 18
    × 9

14. 38
    × 5

15. 26
    × 6

16. 74
    × 8

17. 87
    × 5

18. 53
    × 7

19. 49
    × 3

20. 71
    × 4

Fluency Practice

Multiply.

1. 11
    × 23

2. 54
    × 41

3. 76
    × 15

4. 35
    × 64

5. 27
    × 10

6. 89
    × 33

7. 41
    × 48

8. 92
    × 13

9. 63
    × 25

10. 39
    × 67

11. 89
    × 40

12. 19
    × 84

13. 218
    × 13

14. 104
    × 37

15. 921
    × 26

16. 585
    × 48

17. 732
    × 55

18. 337
    × 79

19. 376
    × 80

20. 890
    × 14

Vocabulary Check

Match each word to its definition. Write your answers on the lines provided.

1. **compatible numbers** _____

2. **product** _____

3. **power of 10** _____

4. **factor** _____

5. **prime factorization** _____

6. **Distributive Property** _____

7. **exponent** _____

8. **base** _____

9. **power** _____

A. This is a number obtained by raising a base to an exponent.

B. This property states that to multiply a number by a sum, you can multiply each addend by the number and add the products.

C. This is a number that divides into a whole number evenly. It is also a number that is multiplied by another number.

D. This is a way of expressing a composite number as a product of its prime factors.

E. This is a number like 10, 100, 1,000, and so on. It is the result of using only 10 as a factor.

F. These are numbers that are easy to multiply mentally.

G. This is the answer to a multiplication problem.

H. In a power, this number represents the number of times the base is used as a factor.

I. In a power, this is the number used as the repeated factor.

Concept Check

Find the prime factorization of each number.

10. $12 =$ _____

11. $42 =$ _____

Write each product using an exponent.

12. $10 \times 10 =$ _____

13. $5 \times 5 \times 5 \times 5 =$ _____

Find each product mentally.

14. $73 \times 10^2 =$ _____

15. $60 \times 40 =$ _____

Estimate. Then multiply. Use your estimate to check your answer.

16.
$$\begin{array}{r} 72 \\ \times\ 36 \\ \hline \end{array}$$

17.
$$\begin{array}{r} 23 \\ \times\ 84 \\ \hline \end{array}$$

18.
$$\begin{array}{r} 321 \\ \times\ 64 \\ \hline \end{array}$$

**Find each product mentally using the Distributive Property.
Show the steps that you used.**

19. $8 \times 71 =$ _____

20. $6 \times 83 =$ _____

Problem Solving

For Exercises 21–23, use the following information. Then estimate to find the distance sound travels through each material in each given time.

Sound travels through different materials at different speeds. For example, the graph shows that in one second, sound travels 5,971 meters through stone. However, it travels only 346 meters through air in one second.

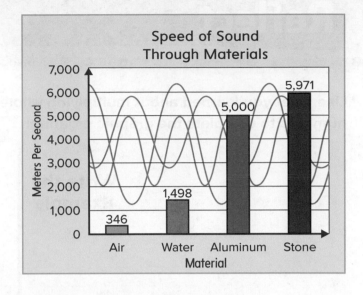

Speed of Sound Through Materials

21. air, 20 seconds

_____

22. stone, 12 seconds

_____

Brain Builders

23. Estimate how much farther sound travels through stone in 17 seconds than through aluminum in the same time. Show your work.

_____

24. Sylvia is saving to buy a new terrarium for her iguana. She saves $2 the first week, $4 the second week, $8 the third week, and so on. How much total money will she save in 5 weeks? Solve by completing the table. If Sylvia wants to buy a terrarium that costs $125, in how many weeks will she have enough money? Explain.

Week	1	2	3	4	5
Amount Saved ($)	2	4	8		

_____

25. Test Practice Colin bought 7 flats of flowers. Each flat contains 24 flowers. How many flowers did he buy?

Ⓐ 140 flowers   Ⓒ 168 flowers

Ⓑ 154 flowers   Ⓓ 200 flowers

Reflect

Use what you learned about multiplying whole numbers to complete the graphic organizer.

Write the Example

Real-World Example

□ □ □

ESSENTIAL QUESTION

What strategies can be used to multiply whole numbers?

Vocabulary

Model

Now reflect on the **ESSENTIAL QUESTION** Write your answer below.

_____

_____

Performance Task

Brain Builders

Buying Cards

Maya is ordering special trading cards for her store. You can play a game with these cards which are modeled after a popular video game. The boxes come with 33 packs and each pack has 9 cards.

Show all your work to receive full credit.

Part A

A customer has requested 11 boxes. Maya estimates the order will be about 300 packs. Is her estimate higher or lower than the actual total? Explain your reasoning.

Part B

Suggest a more accurate way of estimating the number of total packs in 9 boxes. Explain why your estimate will be more accurate.

Part C

The store uses the following area model to find the total number of packs in 11 boxes. Complete the area model by filling in the boxes to determine the total number of packs.

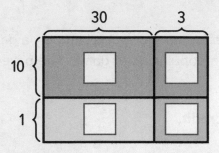

_____ total packs

Part D

How many total cards are in 11 boxes? Explain your reasoning.

Chapter

3 Divide by a One-Digit Divisor

ESSENTIAL QUESTION

What strategies can be used to divide whole numbers?

Let's Help Others!

Watch

Watch a video!

149

MY Chapter Project

Real-World Division

1. Think of times when you have used division in your real-life experiences. Together with your group, brainstorm a list of real-life experiences that use division. List all your experiences below.

2. Choose 5 or more of the experiences listed above and create a poster that illustrates each real-life experience. Include a number model for each experience you illustrate. Show how to interpret the remainder if any of your number models have a remainder.

3. As a group, present your poster to the class.

Am I Ready?

Multiply.

1. 12 × 7 = _____

2. 42 × 8 = _____

3. 51 × 9 = _____

4. 7 × 18 = _____

5. 3 × $75 = _____

6. 3 × $89 = _____

7. Turner's bookshelf has 6 shelves. Each shelf has 17 books. How many books are on the bookshelf?

_____

Round each number to its greatest place value.

8. 36 _____

9. $451 _____

10. 7,499 _____

11. $33,103 _____

12. $271 _____

13. $5,001 _____

14. There are 7,209 students at the amusement park. To the nearest thousand, how many students are at the park?

_____

Divide.

15. 8 ÷ 2 = _____

16. 15 ÷ 3 = _____

17. 27 ÷ 3 = _____

18. 28 ÷ 4 = _____

19. 48 ÷ 6 = _____

20. 54 ÷ 9 = _____

21. Three people spent a total of $24 for lunch. If they divided the total cost equally, how much did each person pay?

_____

Item	Cost
Pizza	$12
Salads	$6
Drinks	$6

Shade the boxes to show the problems you answered correctly.

How Did I Do? | 1 | 2 | 3 | 4 | 5 | 6 | 7 | 8 | 9 | 10 | 11 | 12 | 13 | 14 | 15 | 16 | 17 | 18 | 19 | 20 | 21 |

Name

....

MY Math Words Vocab abc

Review Vocabulary

compatible numbers   multiples   place value   product

Making Connections

Use the review vocabulary to complete the concept wheel.

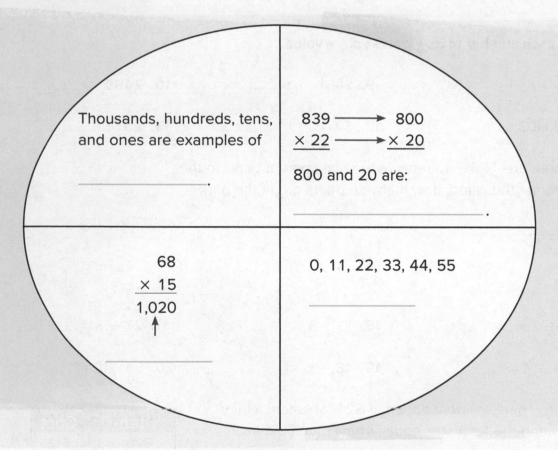

Thousands, hundreds, tens, and ones are examples of

_____ .

839 ⟶ 800
× 22 ⟶ × 20

800 and 20 are:

_____ .

68
× 15
1,020
↑

_____

0, 11, 22, 33, 44, 55

_____

How can these vocabulary words help you solve multiplication problems?

_____

_____

Lesson 3–3

dividend

$$76 \div 4 = 19$$

Lesson 3–3

divisor

$$76 \div 4 = 19$$

Lesson 3–1

fact family

28

$\times, \div$

4   7

$4 \times 7 = 28$
$7 \times 4 = 28$
$28 \div 7 = 4$
$28 \div 4 = 7$

Lesson 3–7

partial quotients

```
  8)536
  −480      60
    56     + 7
  − 56      67
     0
```

Lesson 3–3

quotient

$$76 \div 4 = 19$$

Lesson 3–3

remainder

```
    79 R3
 4)319
  −28
    39
  −36
     3
```

Lesson 3–1

unknown

$42 \div 6 = \blacksquare$

$6 \times 7 = 42$

$42 \div 6 = 7$

Lesson 3–1

variable

$$4 \times k = 32$$

Ideas for Use

- Work with a partner to name the part of speech of each word. Consult a dictionary to check your answers.

- Draw or write examples for each card. Be sure your examples are different from what is shown on each card.

The number that divides the dividend.

Write 3 division problems that have 2-digit dividends and 1-digit divisors. Circle each divisor.

_____

_____

A number that is being divided.

A division problem has a divisor of 12 and a quotient of 7. What is the dividend?

_____

A dividing method in which the dividend is separated into addends that are easy to divide.

How can you use place value when using partial quotients?

_____

_____

A group of related facts that use the same numbers.

Write a set of numbers that are a fact family. Explain why they are a fact family.

_____

_____

The number that is left after one whole number is divided by another.

Write the letter used to represent *remainder*.

_____

The result of a division problem.

Which review word is also the result of an operation on numbers?

_____

A letter or symbol used to represent an unknown quantity.

The Latin root *var* means "different." Explain how this helps you understand the meaning of *variable*.

_____

A missing value.

50 divided by an unknown number is equal to 5. Find the unknown number.

_____

FOLDABLES® Follow the steps on the back to make your Foldable.

My Division Strategies

Quotients with Zeros

Example:

Two-Digit Dividends

Example:

Division Models

Example:

Interpret the Remainder

Example:

Estimate Quotients

Example:

Place the First Digit

Example:

Lesson 1
Relate Division to Multiplication

A **fact family** is a group of related facts that use the same numbers. You can use fact families to relate multiplication and division.

 Math in My World

Example 1

Sheryl is helping to put away 18 basketballs after practice. She places them on a rack that has 3 shelves. How many basketballs can she put on each self?

Use a fact family.

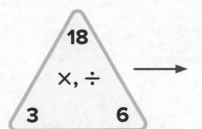

$3 \times \underline{\hspace{1cm}} = 18$

$6 \times \underline{\hspace{1cm}} = 18$

$18 \div 6 = \underline{\hspace{1cm}}$

$18 \div 3 = \underline{\hspace{1cm}}$

So, $18 \div 3 = \underline{\hspace{1cm}}$. Sheryl can put $\underline{\hspace{1cm}}$ basketballs on each shelf.

Check Draw an equal amount of basketballs on each shelf.

$\underline{\hspace{1cm}}$ shelves

$\underline{\hspace{1cm}} \times \underline{\hspace{1cm}} = 18$

$\underline{\hspace{1cm}}$ basketballs on each shelf

 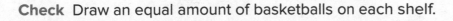

An equation is a number sentence that contains an equals sign (=). You can use related facts to help find the **unknown,** or missing value, in an equation. You can use a **variable,** or a letter, to represent the unknown number.

Example 2

Ellie is creating gift bags for her party guests. She wants to divide 56 pencils equally among the 7 gift bags. How many pencils will go in each bag?

Let p represent the number of pencils in each bag.

_____ ÷ _____ = p

Think: What number times 7 is 56?

Write a related multiplication fact.

$7 \times$ _____ $= 56$

So, $56 \div 7 =$ _____ . Since $p =$ _____ , Ellie will put _____ pencils in each bag.

Guided Practice

1. Complete the fact family for 8, 9, 72.

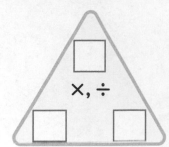

$8 \times$ _____ $= 72$

$9 \times$ _____ $= 72$

_____ $\div 8 =$ _____

_____ $\div 9 =$ _____

Describe how you could use multiplication to find $21 \div 7 = x$.

Divide. Use a related multiplication fact.

2. $48 \div$ _____ $= 6$

Think: _____ $\times 6 = 48$

3. $40 \div 5 =$ _____

Think: $5 \times$ _____ $= 40$

Independent Practice

Write a fact family for each set.

4.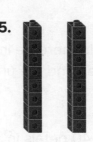

5.

_____   _____   _____

_____   _____   _____

6. 4, 9, 36   **7.** 5, 7, 35   **8.** 3, 8, 24

_____   _____   _____

_____   _____   _____

_____   _____   _____

Divide. Write the related multiplication fact.

9. $64 \div 8 =$ ____   **10.** $45 \div 9 =$ ____   **11.** ____ $\div 9 = 9$

____ $\times 8 = 64$ ____ $\times 9 = 45$ $9 \times 9 =$ ____

12. ____ $\div 8 = 4$ **13.** $40 \div$ ____ $= 8$ **14.** $63 \div$ ____ $= 7$

$4 \times 8 =$ ____ $8 \times$ ____ $= 40$ $7 \times$ ____ $= 63$

Algebra Find the unknown number in each equation.
Use a related division fact.

15. $2 \times m = 12$ **16.** $8 \times y = 24$ **17.** $9 \times g = 72$

$m =$ ____ $y =$ ____ $g =$ ____

Lesson 1 Relate Division to Multiplication   **159**

Problem Solving

Algebra For Exercises 18–20, use the information below.
Orange blossoms have 5 petals and are some of the most fragrant flowers.

18. How many petals would there be in a group of 7 flowers?

19. How many petals *p* would there be in a group of 11 flowers? Write an equation to find the unknown. Then find the unknown.

20. A group of *f* flowers has 40 petals in all. Write an equation to find the unknown. Then find the unknown.

Brain Builders

21. **Processes &Practices**  **Reason** Can the number 12 be part of more than one fact family? Explain and give an example. Make a general statement about fact families that applies to all numbers.

22. **Processes &Practices** **3** **Which One Doesn't Belong?** Circle the equation that does not belong with the other three. Explain why it does not belong.

$$54 \div 9 = 6 \qquad 54 \div 6 = 9 \qquad 9 \times 3 = 27 \qquad 6 \times 9 = 54$$

23. **Building on the Essential Question** How do multiplication facts help me divide? Give an example.

Name

MY Homework

Homework Helper

Need help? ↗ connectED.mcgraw-hill.com

There are 20 students in the after-school program. There are 4 students in each group. How many groups are there?

Use a fact family.

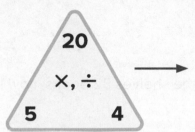

$4 \times 5 = 20$

$5 \times 4 = 20$

$20 \div 4 = 5$

$20 \div 5 = 4$

So, $20 \div 4 = 5$. There are 5 groups in the after-school program.

Check Use a drawing.

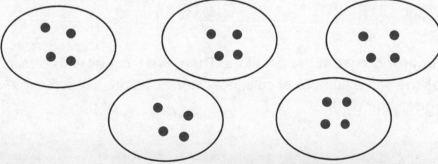

Practice

Write a fact family for each set of numbers.

1. 2, 10, 20

2. 8, 7, 56

3. 7, 9, 63

Divide. Write the related multiplication fact.

4. $36 \div$ _____ $= 6$

$6 \times$ _____ $= 36$

5. $77 \div 11 =$ _____

_____ $\times 11 = 77$

6. _____ $\div 8 = 12$

$12 \times 8 =$ _____

Processes &Practices **Use Algebra** Find the unknown number in each equation. Use a related division fact.

7. $4 \times b = 12$

$b =$ _____

8. $3 \times n = 45$

$n =$ _____

9. $6 \times k = 18$

$k =$ _____

Vocabulary Check

10. Choose the correct word(s) to complete the sentence below. A group of related facts that use the same numbers is called a(n) _____.

 # Problem Solving

11. Julian places 8 books on each shelf of a bookcase. If he shelves 32 books, how many shelves are needed?

_____

Brain Builders

12. There are 15 cats ready for new homes at the pet store. There are 5 cages with cats in the store. If each cage has the same number of cats, how many cats are in each cage? Write two equations to model your work.

_____

_____

13. The leaves of a poison ivy plant are found in clusters of 3. One poison ivy plant has a total of 21 leaves. How many leaf clusters does this plant have? Write two equations to model your work.

_____

_____

14. **Test Practice** A local pet store has a total of 72 fish in 9 tanks. Each tank holds the same number of fish. How many fish are in each tank?

Ⓐ 6 fish

Ⓒ 8 fish

Ⓑ 7 fish

Ⓓ 9 fish

Lesson 2
Hands On
Division Models

Build It   Tools

In art class, three students share **48** markers evenly.
How many markers will each student have?

1 Model 48 using base-ten blocks.

2 Divide the tens into 3 equal groups.
Circle 3 groups of tens. Draw the equal groups.

3 Regroup the remaining tens blocks into 10 ones.

How many ones are there altogether? _____ ones

4 Divide the ones into 3 equal groups. Draw an equal
amount of ones in each group. Each group has

_____ ones.

Each group has _____ ten and _____ ones.

So, 48 ÷ 3 = _____.

Each student will have _____ markers.

Find 56 ÷ 5.

 Model 56 using base-ten blocks.

 Divide the tens into 5 equal groups. Circle the groups of tens.

Divide the ones into 5 equal groups. Draw an equal amount of tens and ones in each group.

How many ones are left over? _____

Each group has _____ ten and _____ one.

So, when you divide 56 into 5 groups, there are _____ in each group with one left over.

Processes &Practices  **Use Math Tools** Model 32 ÷ 3 using base-ten
1. blocks. Will there be any left over? Explain.

Practice It

**Divide. Use base-ten blocks. Draw the equal groups.
State if there are any left over.**

2. 44 ÷ 4

How many are in each group? _____

Are there any left over? If so,
state how many. _____

3. 39 ÷ 3

How many are in each group? _____

Are there any left over? If so, state how
many. _____

4. 32 ÷ 5

How many are in each group? _____

Are there any left over? If so, state how
many. _____

5. 57 ÷ 8

How many are in each group? _____

Are there any left over? If so, state how
many. _____

6. Use base-ten blocks to divide 64 ÷ 6. How many are left over?

_____

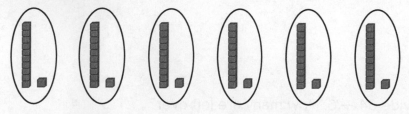

Apply It

Solve. Use base-ten blocks.

7. **Processes &Practices** **5** **Use Math Tools** Kendrick has 42 craft sticks to make 3 identical crafts. How many sticks will he use for each craft?

_____

8. Allison uses 41 stickers to decorate 3 picture frames. Each picture frame has the same number of stickers. How many stickers are on each picture frame? How many stickers are left over?

_____

9. Kara and six of her friends are playing miniature golf. If it costs $42 for all of them to play a round of golf, how much does each round cost per person?

_____

10. Justin has 71 newspapers to deliver in 3 days. He delivers the same number of newspapers each day. How many newspapers does Justin deliver each day? How many newspapers are left over?

_____

_____

11. **Processes &Practices** **2** **Use Number Sense** Write the division sentence that is shown by the model.

_____

Write About It

12. How does place value help model division?

_____

_____

MY Homework

Homework Helper eHelp

Need help? connectED.mcgraw-hill.com

Find 71 ÷ 6.

1 Model 71 using base-ten blocks.

2 Divide the tens into 6 equal groups. Circle 6 groups of ten.

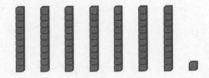

3 Regroup the remaining tens blocks into 10 ones.
You now have a total of 11 ones.

4 Divide the ones into 6 equal groups. Draw an equal
amount of tens and ones in each group.

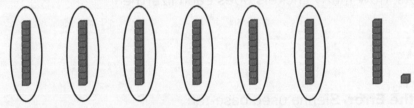

There are 5 ones left over.

So, when you divide 71 into 6 groups, there are
11 in each group with five left over.

Practice

**Divide. Use base-ten blocks. Draw the equal groups.
State if there are any left over.**

1. $42 \div 3$

   How many are in each group? _____

   Are there any left over? _____

2. $87 \div 4$

   How many are in each group? _____

   Are there any left over? _____

Problem Solving

3. Jamil has 3 pet lizards. The pet store owner said that Jamil will need to buy a total of 36 crickets to feed his lizards. If each lizard eats the same number of crickets, how many crickets does each lizard eat?

   _____

4. **Processes &Practices** **3** **Find the Error** Sienna used base-ten blocks to find $45 \div 3$. Explain her error.

   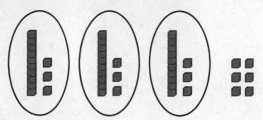

   _____

   _____

5. An airplane that can hold 63 passengers is separated into 3 sections. Each section holds the same number of passengers. Write and solve a division sentence to correctly describe the situation.

   _____

Lesson 3
Two-Digit Dividends

ESSENTIAL QUESTION
What strategies can be used to divide whole numbers?

The **dividend** is the number that is being divided.
The **divisor** tells you how many groups.

$$36 \div 3 \xleftarrow{\text{divisor}} \quad 3\overline{)36}$$

dividend

The result of the division is called the **quotient.**

 Math in My World | Watch | Tools | Tutor

Example 1

Eli donates his toys to 5 different charities. He has a total of 75 toys to donate. Eli donates the same number of toys to each charity. How many toys does each charity receive?

Let *t* represent the number of toys each charity receives.

_____ ÷ _____ = t

Find 75 ÷ 5.

1 **Divide the tens.** 7 ÷ 5
Write 1 in the quotient over the tens place.

2 **Multiply.** 5 × 1
Subtract. 7 − 5

3 **Bring down the ones.**

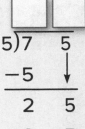

$$5\overline{)7 \quad 5}$$
$$-5$$
$$2 \quad 5$$
$$-2 \quad 5$$
$$0$$

4 **Divide the ones.**
25 ÷ 5
Write 5 in the quotient over the ones place.

5 **Multiply.** 5 × 5
Subtract. 25 − 25

The model shows 5 groups of fifteen.

So, 75 ÷ 5 = _____.

Each charity receives _____ toys.

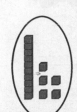    

A **remainder** is the number, or part left, after you divide.
Use R to represent the remainder.

Example 2

**Caleb is putting his baseball cards in an album. He has
57 cards and can put 4 cards on each page. How many
full pages will Caleb have? Will there be any cards left?**

Find 57 ÷ 4.

 Divide the tens.

5 ÷ 4

Write 1 in the quotient over
the tens place.

 Multiply. 4 × 1
Subtract. 5 − 4
Compare. 1 < 4

 Bring down the ones.

$$4\overline{)5\quad7}\qquad\boxed{}\boxed{}\ \text{R}\ \boxed{}$$

 Divide the ones.

17 ÷ 4

Write 4 in the quotient
over the ones place.

 Multiply. 4 × 4
Subtract. 17 − 16
Compare. 1 < 4
The remainder is _____.

So, there will be _____ full pages and _____ card will be left over.

Guided Practice

1.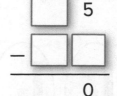

$$5\overline{)6\quad5}$$
$$-\ 5$$
$$\boxed{}\ 5$$
$$-\ \boxed{}\boxed{}$$
$$0$$

2.

$$3\overline{)4\quad5}$$
$$-\ \boxed{}$$
$$\boxed{}\ 5$$
$$-\ 1\ \boxed{}$$
$$0$$

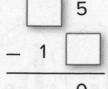

Talk MATH

What should you do if the
remainder is greater than or
equal to the divisor?

Independent Practice

Divide.

3.

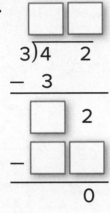

$$\begin{array}{r} \square\ \square \\ 3\overline{)4\ \ 2} \\ -\ 3 \\ \hline \square\ 2 \\ -\ \square\ \square \\ \hline 0 \end{array}$$

4.

$$\begin{array}{r} \square\ \ 7 \\ 4\overline{)6\ \ 8} \\ -\ \square \\ \hline \square\ \ 8 \\ -\ 2\ \square \\ \hline 0 \end{array}$$

5.

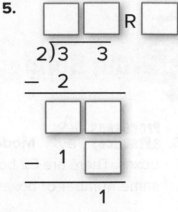

$$\begin{array}{r} \square\ \square\ R\ \square \\ 2\overline{)3\ \ 3} \\ -\ 2 \\ \hline \square\ \square \\ -\ 1\ \square \\ \hline 1 \end{array}$$

6. $2\overline{)28}$

7. $6\overline{)74}$

8. $7\overline{)85}$

9. $60 \div 4 =$ _____

10. $64 \div 5 =$ _____

11. $70 \div 6 =$ _____

Algebra Divide to find the unknown number in each equation.

12. $48 \div 3 = h$

$h =$ _____

13. $44 \div 2 = b$

$b =$ _____

14. $72 \div 4 = w$

$w =$ _____

15. Maranda practiced a total of 52 hours in 4 weeks to prepare for a piano recital. If she practiced the same number of hours each week, how many hours did she practice each week?

_____

Brain Builders

16. **Processes &Practices** **4** **Model Math** Five students volunteered to carry boxes. There are 62 boxes. Is it possible for each student to carry the same number of boxes and have all the boxes carried? Explain.

_____

_____

17. **Processes &Practices** **2** **Reason** The following equations show the relationship between multiplication and division.

$$18 \div 3 = 6 \qquad\qquad 3 \times 6 = 18$$

$$18 \div 0 = ? \qquad\qquad 0 \times ? = 18$$

Explain why it is not possible to divide by zero. Write a true multiplication equation to support your explanation.

_____

_____

_____

18. **Building on the Essential Question** How does place value help me divide? Draw a model to support your explanation.

_____

_____

_____

MY Homework

Homework Helper

Need help? connectED.mcgraw-hill.com

Find 87 ÷ 6.

 Divide the tens.

$8 ÷ 6$

Write 1 in the quotient over the tens place.

$$\begin{array}{r} 14\ R3 \\ 6\overline{)87} \\ -\ 6\!\downarrow \\ \hline 27 \\ -24 \\ \hline 3 \end{array}$$

 Divide the ones.

$27 ÷ 6$

Write 4 in the quotient over the ones place.

2 **Multiply.** $6 × 1$
Subtract. $8 − 6$
Compare. $2 < 6$

5 **Multiply.** $6 × 4$
Subtract. $27 − 24$
Compare. $3 < 6$
The remainder is 3.

3 Bring down the ones.

Practice

Divide.

1. $3\overline{)63}$

2. $7\overline{)96}$

3. $5\overline{)68}$

Algebra Divide to find the unknown number in each equation.

4. $72 ÷ 6 = n$

$n =$ _____

5. $45 ÷ 3 = p$

$p =$ _____

6. $52 ÷ 2 = k$

$k =$ _____

7. A book has 5 chapters and a total of 90 pages. If each chapter has the same number of pages, how many pages are in each chapter?

_____

8. **Processes &Practices** ➋ **Reason** Caitlin divides a bag of fruit snacks with 4 friends. She divides the 89 snacks equally. How many fruit snacks does each person receive? How many fruit snacks will be left over?

_____

Brain Builders

9. Chet is camping with 75 people from his after-school club. A tent can hold up to 8 people. How many tents are needed? Explain your thinking.

_____

_____

10. Isaiah helped pick 72 bananas on the weekend. There were a total of 6 people picking bananas. If they each picked an equal number of bananas, how many bananas did each person pick? Write a multiplication and division equation to support your answer.

_____

_____

11. **Test Practice** A box of granola bars has 26 bars. If 7 friends split the bars equally, how many bars will be left?

   Ⓐ 2 bars   Ⓒ 4 bars

   Ⓑ 3 bars   Ⓓ 5 bars

Lesson 4
Division Patterns

ESSENTIAL QUESTION
What strategies can be used to divide whole numbers?

You can use basic facts and patterns to divide multiples of 10.

 Math in My World Watch ▶ Tutor 💬

Example 1

A monarch butterfly can fly 240 miles in 3 days. Suppose it flies the same distance each day. How many miles can it fly each day?

Find 240 ÷ 3.

1 Since 240 is a multiple of 10, use the basic fact and continue the pattern.

$$2 \quad 4 \div 3 \qquad \text{basic fact}$$

$$2 \quad 4 \;\boxed{}\; \div 3 \qquad \text{24 tens divided by 3 equals 8 tens.}$$

$$2, \; 4 \;\boxed{}\;\boxed{}\; \div 3 \qquad \text{24 hundreds divided by 3 equals 8 hundreds.}$$

$$2 \; 4, \;\boxed{}\;\boxed{}\;\boxed{}\; \div 3 \qquad \text{24 thousands divided by 3 equals 8 thousands.}$$

2 Circle the pattern above that matches the problem.

So, the butterfly can fly _____ miles each day.

Example 2

A cow eats 900 pounds of hay over a period of 30 days. How many pounds of hay would the cow eat each day at this rate?

Let h represent the number of pounds of hay.

_____ ÷ _____ = h

One Way Use a fact family.

Use the fact family of 3, 3, and 9 to help represent the problem.

$3 \times 3 = 9$ ⟷ $9 \div 3 = 3$

$3\,\boxed{} \times 3 = 90$ ⟷ $9\,\boxed{} \div 3\,\boxed{} = 3$

$3\,\boxed{} \times 3\,\boxed{} = 900$ ⟷ $9\,\boxed{}\boxed{} \div 3\,\boxed{} = 3\,\boxed{}$

Another Way Use a pattern in the number of zeros.

You can cross out the same number of zeros in the dividend and the divisor to make the division easier.

$900 \div 30$ Cross out the same number of zeros in both
                              the dividend and divisor.

$90 \div 3 = 30$ Divide. THINK: 9 tens ÷ 3 = 3 tens.

So, $900 \div 30 =$ _____. Since $h =$ _____ , the cow eats

_____ pounds each day.

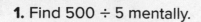
Guided Practice

1. Find $500 \div 5$ mentally.

$500 \div 5 =$ _____

$5 \div 5 = 1$

$5\,\boxed{} \div 5 = 1\,\boxed{}$

$5\,\boxed{}\boxed{} \div 5 = 1\,\boxed{}\boxed{}$

Is the quotient $48 \div 6$ equal to the quotient $480 \div 60$? Explain.

Name _____

Independent Practice

Divide mentally.

2. $800 \div 2 =$ _____

3. $900 \div 3 =$ _____

4. $150 \div 5 =$ _____

5. $140 \div 7 =$ _____

6. $450 \div 9 =$ _____

7. $280 \div 4 =$ _____

8. $180 \div 60 =$ _____

9. $240 \div 30 =$ _____

10. $420 \div 70 =$ _____

Algebra Mentally find each unknown.

11. $1{,}800 \div 30 = k$

$k =$ _____

12. $2{,}000 \div 400 = z$

$z =$ _____

13. $2{,}400 \div 300 = s$

$s =$ _____

Problem Solving

14. **Processes &Practices** **Make Sense of Problems** A group of 10 people bought tickets to a reptile exhibit and paid a total of $130. What was the price of one ticket?

_____

15. The fastest team in a wheelbarrow race traveled 100 meters in about 20 seconds. On average, how many meters did the team travel each second?

_____

16. A video store took in $450 in DVD rentals during one day. If DVDs rent for $9 each, how many DVDs were rented?

_____

Brain Builders

17. Daniela has a 162-ounce bag of potting soil. She puts an equal amount of soil in 4 pots. How much soil will she put in each pot? How many ounces are left over? Explain how you know.

_____

18. **Processes &Practices** **3** **Find the Error** Sonia is finding $5{,}400 \div 90$ mentally. Find her mistake. Tell how Sonia can use place value concepts to avoid this mistake.

$$5{,}4\cancel{0}\cancel{0} \div 9\cancel{0}$$
$$54 \div 9 = 6$$

_____

19. **Building on the Essential Question** How can patterns help me divide multiples of 10? Explain.

_____

Name

MY Homework

Homework Helper

Need help? connectED.mcgraw-hill.com

Find 630 ÷ 7.

 Since 630 is a multiple of 10, use the basic fact and continue the pattern.

$$63 \div 7 \qquad \text{basic fact}$$

$$\boxed{630 \div 7} \qquad \text{63 tens divided by 7 equals 9 tens.}$$

$$6,300 \div 7 \qquad \text{63 hundreds divided by 7 equals 9 hundreds.}$$

$$63,000 \div 7 \qquad \text{63 thousands divided by 7 equals 9 thousands.}$$

Circle the pattern above that matches the problem.

So, 630 ÷ 7 = 90.

Practice

Divide mentally.

1. 270 ÷ 3 = _____

2. 3,200 ÷ 80 = _____

3. 320 ÷ 8 = _____

Algebra Mentally find each unknown.

4. 2,000 ÷ 10 = m

5. 8,100 ÷ 90 = b

6. 450 ÷ 9 = r

m = _____

b = _____

r = _____

Problem Solving

Processes
7. **&Practices** ➊ **Make a Plan** Peyton has collected 120 aluminum cans for recycling. If 20 cans will fit in each blue plastic bag, how many bags will she need to carry all the cans?

_____

8. There are 5,000 sheets of paper in 25 boxes. If each box contains the same number of sheets of paper, how many sheets of paper are in each box?

_____

9. Richard measured his rectangular living room. The room has an area of 200 square feet, and the length is 20 feet. The width is found by dividing the area by the length. What is the width of Richard's living room?

_____

Brain Builders

10. The average polar bear weighs 1,200 pounds. The average grizzly bear weighs 800 pounds. The average black bear weighs 400 pounds. The average polar bear weighs how many times more than the average black bear?

_____

The average grizzly bear weighs how many times more than the average black bear?

_____

11. **Test Practice** An elementary school has 320 students. All of the students are going on a field trip. If 40 students can ride a bus, how many buses are needed?

(A) 6 buses   (C) 8 buses

(B) 7 buses   (D) 9 buses

Check My Progress

Vocabulary Check

Draw lines to match each definition to each vocabulary word.

1. the number that divides the dividend • **fact family**

2. a group of related facts using the same numbers • **variable**

3. a missing value in a number sentence or equation • **divisor**

4. a number that is being divided • **quotient**

5. the result of a division problem • **dividend**

6. a letter or symbol used to represent an unknown quantity • **remainder**

7. the number that is left after one whole number is divided by another • **unknown**

Concept Check

Divide. Write the related multiplication fact.

8. $54 \div 9 =$ _____

_____ $\times 9 = 54$

9. _____ $\div 9 = 8$

$8 \times 9 =$ _____

Algebra Divide to find the unknown number in each equation.

10. $95 \div 5 = n$

$n =$ _____

11. $96 \div 8 = b$

$b =$ _____

Divide.

12. $2\overline{)48}$

13. $7\overline{)81}$

Divide mentally.

14. $3,500 \div 5 =$ _____

15. $420 \div 60 =$ _____

Problem Solving

16. Suki received $87 for working 3 days. If she made the same amount each day, how much did Suki earn each day?

_____

Brain Builders

17. A total of 180 students went on a field trip. There were 3 buses. If each bus had the same number of students on it, how many students were on each bus? Explain how to solve using compatible numbers.

_____

_____

18. Marc is helping out with the school bake sale. He has 50 cookies to place in bags. He places 3 cookies in each bag. How many bags will he use? How many cookies will be left over?

_____

19. **Test Practice** A train traveled 300 miles in 5 hours. How far did the train travel each hour, on average?

Ⓐ 60 miles   Ⓒ 600 miles

Ⓑ 150 miles   Ⓓ 1,500 miles

Lesson 5
Estimate Quotients

To estimate a quotient, you can use compatible numbers, or numbers that are easy to divide mentally. Look for numbers that are part of fact families.

 Math in My World Watch Tutor

Example 1

A dog's heart beats 365 times in 3 minutes. About how many times does a dog's heart beat in 1 minute?

Estimate 365 ÷ 3.

 Change 365 to 360 because 360 and 3 are compatible numbers.

$$365 \quad \div 3$$

$$\boxed{\ }\boxed{\ }\boxed{\ } \div 3 = \boxed{\ }\boxed{\ }\boxed{\ }$$

 Divide mentally.

So, a dog's heart beats about _____ times a minute.

Check Multiply to check your answer.

_____ × 3 = _____

You can use both rounding and compatible numbers to help estimate.

Example 2

Estimate 208 ÷ 8.

1 Round the dividend to the nearest hundred.

2 Change the divisor to a number that is compatible with the rounded dividend.

3 Divide mentally.

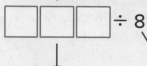

So, 208 ÷ 8 is about _____ .

Guided Practice

1. Estimate 850 ÷ 9.

Change 850 to 900 because 900 and 9 are compatible numbers.

$$850 \div 9$$

Divide mentally.

So, 850 ÷ 9 is about _____ .

Talk MATH

Explain how you could use compatible numbers to estimate 272 ÷ 4.

Independent Practice

Estimate. Show how you estimated.

2. $635 \div 8$

$\downarrow$

_____ $\div 8 =$ _____

3. $545 \div 5$

$\downarrow$

_____ $\div 5 =$ _____

4. $431 \div 2$

$\downarrow$

_____ $\div 2 =$ _____

5. $374 \div 9$

6. $395 \div 4$

7. $660 \div 7$

8. $289 \div 9$

9. $477 \div 9$

10. $230 \div 7$

11. $639 \div 7$

12. $350 \div 8$

13. $584 \div 6$

Problem Solving

Processes &Practices **5** **Use Math Tools** A grocery store employee puts 8 bagels in each bag. If she has 385 bagels, about how many bags does she need?

14.

15. A student email to troops contained 250 characters. The 4-line email contained the same number of characters on each line. About how many characters were on each line? Show how you estimated.

16. Jani drives 290 miles in 5 hours. About how many miles does she drive each hour?

Brain Builders

17. **Processes &Practices** **2** **Use Number Sense** Write a division problem. Show two different ways that you can estimate the quotient using compatible numbers. Which estimate is best? Explain why you think so.

18. **Building on the Essential Question** Why is it important to know how to estimate quotients? Give 2 reasons.

Name

MY Homework →

Homework Helper

Need help? connectED.mcgraw-hill.com

Estimate 488 ÷ 9.

1 Round the dividend to the nearest hundred.

2 Change the divisor to a number that is compatible with the rounded dividend.

3 Divide mentally.

So, 488 ÷ 9 is about 50.

$$488 \div 9$$
$$\downarrow$$
$$500 \div 9$$
$$\downarrow \qquad \downarrow$$
$$500 \div 10 = 50$$

Practice

Estimate. Show how you estimated.

1. 115 ÷ 2

2. 791 ÷ 2

3. 151 ÷ 3

4. 460 ÷ 9

5. 477 ÷ 7

6. 392 ÷ 9

Problem Solving

7. The table shows how much each fifth grade room earned from a bake sale. The money is going to be given to 6 different charities. If each charity is given an equal amount, about how much will each charity receive? Show how you estimated.

Bake Sale

Room	Earnings($)
110	327
112	425
114	550
116	486

_____

_____

_____

8. There were 317 marbles divided equally among 8 bowls. About how many marbles were in each bowl?

_____

_____

Brain Builders

9. **Processes &Practices** 5 **Use Math Tools** Emilio has 5 bags of birdseed. Each bag has about 28 ounces of birdseed. If he divides the birdseed equally into 3 containers, about how much birdseed will he put in each container? Model your work using equations.

_____

_____

10. **Test Practice** Which of the following is the most reasonable estimate for the number of Calories in one serving of milk? In one-half serving of milk?

Ⓐ between 8 and 9 ; between 16 and 18

Ⓑ less than 80 ; less than 40

Ⓒ between 80 and 90 ; between 40 and 45

Ⓓ more than 90 ; more than 45

Servings of Milk	Calories
5	430

Lesson 6
Hands On
Division Models with Greater Numbers

Build It Tools

At the fair, you need tickets to ride the rides. Three friends share 336 tickets equally. How many tickets will each friend receive?

Find 336 ÷ 3.

1 Model 336 using base-ten blocks.

2 Divide the hundreds into 3 groups. Circle the equal groups above.

How many hundreds are in each group? _____

3 Divide the tens into 3 groups. Circle the equal groups above.

How many tens are in each group? _____

4 Divide the ones into 3 groups. Circle the equal groups above.

How many ones are in each group? _____

    1   1   2

hundred   ten   ones

_____ + _____ + _____ = _____

So, each friend will receive _____ tickets.

Check Use multiplication to check your answer.

_____ × 3 = 336

Try It

Find 319 ÷ 2.

The base-ten blocks show 319.

 Divide the hundreds into 2 groups.

Draw the equal groups.

How many hundreds are left over? _____

 Regroup the remaining hundred into

_____ tens.

There are now a total of _____ tens.

Divide the tens into 2 groups.

Draw _____ tens in each group.

How many tens are left over? _____

 Regroup the remaining ten into _____ ones.

There are now a total of _____ ones.
Divide the ones into 2 groups.

Draw _____ ones in each group.

How many ones are left over? _____

Each group has _____ hundred, _____ tens, and _____ ones.

There is _____ one left over. So, 319 ÷ 2 = _____ .

Check Use multiplication to check your answer.

_____ × 2 = 318   318 + _____ = 319

Talk About It

1. **Processes &Practices** **Plan Your Solution** How can you divide 3 hundreds into 2 equal groups?

_____

_____

Practice It

Use models to find each quotient. Draw the equal groups.

2. $344 \div 2 =$ _____

3. $468 \div 4 =$ _____

4. $383 \div 3 =$ _____

5. $257 \div 2 =$ _____

Apply It

For Exercises 6–7 and 9, use models to help you find each quotient.

6. Sarah ordered 366 school newspapers to be printed. The newspapers arrived in 3 boxes. How many newspapers were in each box?

_____

_____

7. A monkey's heart beats about 576 times in 3 minutes. How many times does a monkey's heart beat in 1 minute?

_____

_____

8. **Processes &Practices** 6 **Explain to a Friend** If you are dividing a three-digit even number by two, will you ever have a remainder? Explain to a friend.

_____

_____

9. Gretchen spent 236 hours helping out the neighbors on their farm over the past 2 months. She helped out the same number of hours each month. How many hours did she help each month?

_____

_____

Write About It

10. How can I use models to help me divide?

_____

_____

_____

192 **Chapter 3** Divide by a One-Digit Divisor

MY Homework

Homework Helper

Need help? connectED.mcgraw-hill.com

At the arcade, Jamie and her friends won 476 tickets playing games. The four friends share 476 tickets equally. How many tickets will each friend receive? Find 476 ÷ 4.

1 Model 476 using base-ten blocks.

2 Divide the hundreds into 4 groups.

How many hundreds are in each group? 1

Group 1

3 There are a total of 7 tens.

Divide the tens into 4 groups.

How many tens are in each group? 1

How many tens are left over? 3

Group 2

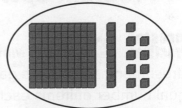

4 Regroup the remaining tens into 30 ones.

There are a total of 36 ones.
Divide the ones into 4 groups.

How many ones are in each group? 9

Group 3

Each group has 1 hundred, 1 ten, and 9 ones.

So, each friend will receive 119 tickets.

Check Use multiplication to check your answer.

119 × 4 = 476

Group 4

Practice

Use models to find each quotient. Draw the equal groups.

1. 328 ÷ 2 = _____

2. 443 ÷ 3 = _____

 **Problem Solving**

3. **Processes &Practices** 4 **Model Math** Denny volunteered to sell 372 refreshments at a sports game. He helped for 3 hours. If he sold an equal number of items each hour, how many items did he sell each hour? Draw models to help you find the quotient.

_____

4. Faith earned $565 last week. She earned the same amount each day. If she worked 5 days, how much did she earn each day? Draw models to help you find the quotient.

_____

Lesson 7
Hands On
Distributive Property and Partial Quotients

ESSENTIAL QUESTION
What strategies can be used to divide whole numbers?

The Distributive Property allows you to divide each place-value position by the same factor.

Draw It

Jesse has 369 beads to be split evenly among 3 necklaces. How many beads can Jesse put on each necklace?

Find 369 ÷ 3.

1 Model 369 as (300 + _____ + _____).

	300		60	9

2 Divide each section by 3.

Write each quotient above the bar.

300 ÷ 3 = _____

60 ÷ 3 = _____

9 ÷ 3 = _____

3	300		60	9

3 Add the quotients.

_____ + _____ + _____ = _____

So, 369 ÷ 3 = _____ .

Jesse can put _____ beads on each necklace.

Check Multiply to check your answer.

_____ × 3 = 369

Partial quotients is a method of dividing where you break the dividend into addends that are easy to divide.

Try It

There are 738 students in Manuel's school. There are 6 grade levels in his school, with each grade having the same number of students. How many students are in the fifth grade?

Find 738 ÷ 6.

1 Divide the hundreds.

600 is close to 738 and is compatible with 6.
Divide 600 by 6.

_____ is a partial quotient.
Subtract 600 from 738.

2 Divide the tens.

120 is close to 138 and is compatible with 6.
Divide 120 by 6.

_____ is a partial quotient.
Subtract 120 from 138.

3 Divide the ones.

Divide 18 by 6.

_____ is a partial quotient.

4 Add the partial quotients.

_____ + _____ + _____ = _____

There are _____ students in the fifth grade.

Check Multiply to check your answer.

_____ × 6 = 738

Partial Quotients

```
    _____
6 ) 7 3 8      _____
  −  6 0 0
  ─────────
     1 3 8
  −  1 2 0      _____
  ─────────
        1 8
  −     1 8      _____
  ─────────
           0
```

Talk About It

1. **Processes &Practices** **Draw a Conclusion** How would you use the Distributive Property to find 242 ÷ 2?

_____

_____

Name ....

Divide. Use the Distributive Property to draw bar diagrams.

2. 248 ÷ 2 = _____

3. 963 ÷ 3 = _____

4. 585 ÷ 5 = _____

5. 488 ÷ 4 = _____

Divide. Use partial quotients.

6. 654 ÷ 3 = _____

7. 675 ÷ 5 = _____

8. 351 ÷ 3 = _____

Apply It

9. Joshua is saving his money to donate to his favorite charity. He has saved $432 so far. If he has been saving for 3 years, how much did he save each year?

_____

10. Mr. Keaton planted 918 cornstalks. There are 9 cornstalks in each row. How many rows of corn did he plant?

_____

11. **Processes &Practices** **2** **Use Number Sense** Colton wants to buy a trampoline in 2 months. The trampoline costs $228. If he saves the same amount each month, how much will he have to save each month?

_____

12. **Processes &Practices** **2** **Reason** Suppose you are finding 296 ÷ 4 using partial quotients. Is 50 or 70 a more reasonable partial quotient? Explain.

_____

_____

Write About It

13. How can properties help me divide?

_____

_____

_____

MY Homework

Homework Helper

Need help? ⤺ connectED.mcgraw-hill.com

The Statue of Liberty has a total of 354 steps. Janice decides to climb the steps in 2 equal sections. How many steps will she climb in each section?

Find 354 ÷ 2.

1 Model 354 as (300 + 50 + 4).

300	50	4

2 Divide each section by 2.
Write each quotient above the bar.
300 ÷ 2 = 150
50 ÷ 2 = 25
4 ÷ 2 = 2

		150	25	2
2	300		50	4

3 Add the quotients.
150 + 25 + 2 = 177

So, 354 ÷ 2 = 177. Janice will climb 177 steps in each section.

Check Multiply to check your answer.   177 × 2 = 354

Practice

1. Divide 844 ÷ 4. Use the Distributive Property to draw a bar diagram.

Problem Solving

2. The fourth, fifth, and sixth grade classes of 396 students take a field trip to the historical museum over 3 days. How many students can go each day?

_____

3. Colton wants to buy a video game system in 5 months. The video game system costs $275. If he saves the same amount each month, how much will he have to save each month?

_____

4. Paul's MP3 player has a total of 852 songs. He has 4 different groups of songs. If he puts an equal number of songs in each group, how many songs will be in each group?

_____

Vocabulary Check

5. Choose the correct word(s) to complete the sentence. Using partial quotients is a method of dividing where you break the

_____ into sections that are easy to divide.

6. **Processes &Practices** ▶**5** **Use Math Tools** Stanley is using the bar diagram to help him find 482 ÷ 2. What is the quotient?

2	400	80	2

_____

Name ....

Lesson 8
Divide Three- and Four-Digit Dividends

ESSENTIAL QUESTION
What strategies can be used to divide whole numbers?

To divide a greater dividend, use the same process as dividing a two-digit dividend.

 Math in My World Watch Tutor

Example 1

In a 4-hour period, 852 people rode an amusement park ride. If the same number of people rode the ride each hour, how many people rode the ride in the first hour?

Let p represent the number of people.

_____ ÷ _____ = p

Find 852 ÷ 4.

Estimate 900 ÷ 4 = _____

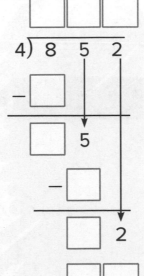

1 Divide the hundreds.
Divide. 8 ÷ 4

Multiply. 2 × 4

Subtract. 8 − 8

Compare. 0 < 4

2 Divide the tens.
Divide. 5 ÷ 4

Multiply. 1 × 4

Subtract. 5 − 4

Compare. 1 < 4

3 Divide the ones.
Divide. 12 ÷ 4

Multiply. 3 × 4

Subtract. 12 − 12

Compare. 0 < 4

So, _____ people rode the ride in the first hour.

Check Multiply to check your answer.

_____ × 4 = 852

 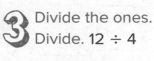

Example 2

Find $6\overline{)7{,}946}$.

1 Divide the thousands.

Divide. $7 \div 6$

Multiply. 1×6

Subtract. $7 - 6$

Compare. $1 < 6$

2 Divide the hundreds.

Divide. $19 \div 6$

Multiply. 3×6

Subtract. $19 - 18$

Compare. $1 < 6$

So, $7{,}946 \div 6 =$ _____.

Check To check division with a remainder, first multiply the quotient and the divisor. Then add the remainder.

_____ $\times 6 = 7{,}944 \longrightarrow 7{,}944 +$ _____ $= 7{,}946$

$\square , \square\square\square$ R $\square$

$6\overline{)7{,} \quad 9 \quad 4 \quad 6}$

$-\square$

$\square \quad 9$

$-\square\square$

$\square \quad 4$

$-\square\square$

$\square \quad 6$

$-\square\square$

$\square$

3 Divide the tens.

Divide. $14 \div 6$

Multiply. 2×6

Subtract. $14 - 12$

Compare. $2 < 6$

4 Divide the ones.

Divide. $26 \div 6$

Multiply. 4×6

Subtract. $26 - 24$

Compare. $2 < 6$

Guided Practice

1. Divide.

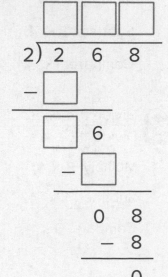

$\square\square\square$

$2\overline{)2 \quad 6 \quad 8}$

$-\square$

$\square \quad 6$

$-\square$

$0 \quad 8$

-8

0

Talk MATH

Does the quotient of 945 and 8 have two or three digits? Explain.

Independent Practice

Divide.

2. 5)595

3. 4)625

4. 5)5,815

5. 516 ÷ 3 = _____

6. 6,418 ÷ 3 = _____

7. 9,345 ÷ 7 = _____

8. 5)755

9. 4)8,468

10. 2)2,349

Algebra Divide to find the unknown number in each equation.

11. 414 ÷ 3 = c

12. 5,120 ÷ 4 = m

13. 1,535 ÷ 5 = x

c = _____

m = _____

x = _____

Problem Solving

14. Three new video game systems cost $645. If all the game systems cost the same, what is the cost of each game system?

_____

15. **Processes &Practices**  **2** **Use Algebra** A state park has cable cars that travel about 864 yards in 4 minutes. The cars travel the same amount of yards each minute. How many yards do the cars travel per minute? Use the bar diagram to write an equation to find the unknown. Then find the unknown.

```
|----------- 864 yards -----------|
| 1 min | 1 min | 1 min | 1 min |
|--? yd--|
```

_____

_____

16. Three adult kangaroos weigh 435 pounds. If each adult weighs the same, how much would one adult kangaroo weigh?

_____

Brain Builders

17. **Processes &Practices** **2** **Use Number Sense** Place the digits 2, 4, 7, and 8 in ■ ■ ■ ÷ ■ to create a division problem with the greatest quotient.

_____

18. **Building on the Essential Question** How can I divide larger dividends? Write a division expression and explain to a friend how to solve.

_____

_____

Name _____

MY Homework

Homework Helper

Need help? ✐ **connectED.mcgraw-hill.com**

Find 630 ÷ 5.

Estimate 600 ÷ 5 = 120

1 Divide the hundreds.

Divide. 6 ÷ 5

Multiply. 1 × 5

Subtract. 6 − 5

Compare. 1 < 5

```
      126
   5)630
    − 5↓
    ─────
       13
     − 10↓
     ─────
        30
      − 30
      ─────
         0
```

2 Divide the tens.

Divide. 13 ÷ 5

Multiply. 2 × 5

Subtract. 13 − 10

Compare. 3 < 5

3 Divide the ones.

Divide. 30 ÷ 5

Multiply. 6 × 5

Subtract. 30 − 30

Compare. 0 < 5

So, 630 ÷ 5 = 126.

Check Multiply to check your answer.   126 × 5 = 630

Practice

Divide.

1. 3)945

2. 3)493

3. 6,315 ÷ 5 = _____

Problem Solving

4. **Processes &Practices** **2** **Use Number Sense** Mr. Peters has 80 sheets of colored paper. Seven of his students need the paper for a project. How many sheets does each student get? How many sheets are left over?

_____

5. Tina has earned a total of 9,644 frequent flyer miles by traveling between Twin Falls and Preston. She has made this trip 4 times. How many miles is one trip between these two cities?

_____

Brain Builders

6. A family of 4 spent $104 for tickets to a concert on Friday and they spent $140 for tickets to a dinner theater on Saturday. All of the tickets were the same price each day. What was the cost of each ticket on Friday and on Saturday?

_____

7. On Monday, a concession stand manager ordered 985 popcorn bags and 1,010 paper food trays. She splits the bags and trays evenly among 5 concession stands. How many items will each concession stand receive? Explain.

_____

8. **Test Practice** A great white shark weighed 4,302 pounds. This weight was 3 times the weight of a blue marlin fish. What was the weight of the blue marlin?

Ⓐ 1,402 pounds   Ⓒ 1,434 pounds

Ⓑ 1,424 pounds   Ⓓ 1,502 pounds

Check My Progress

Vocabulary Check

1. Circle the method that correctly uses **partial quotients**.

```
  4)644  ¦
  -640   ¦  140
  ─────  ¦
    4    ¦
  - 4    ¦   4
  ─────  ¦
    0    ¦
```

```
  3)513  ¦
  -300   ¦  100
  ─────  ¦
   210   ¦
  -150   ¦   50
  ─────  ¦
    60   ¦
  - 60   ¦   20
  ─────  ¦
     0   ¦
```

```
  4)528  ¦
  -400   ¦  100
  ─────  ¦
   128   ¦
  -120   ¦   30
  ─────  ¦
     8   ¦
  -  8   ¦    2
  ─────  ¦
     0   ¦
```

Concept Check

Estimate. Show how you estimated.

2. 244 ÷ 8

3. 700 ÷ 6

4. 890 ÷ 4

Divide.

5. 5)630

6. 1,766 ÷ 6 = _____

7. 2)87

8. Each of the 9 parking lots at an automobile plant holds the same number of new cars. The lots are full. If there are 431 cars in the lots, about how many cars are in each lot? Show how you estimated.

9. A total of 176 valves were used for 8 cars as they were being assembled. The same number of valves were used for each car. How many valves were used for each car?

Brain Builders

10. A construction company estimates that it will take 852 hours to complete a remodeling project. If there are 6 employees that each work an equal number of hours, how many hours will each employee have to work? Can you get the same quotient if you divide 852 by 3 and then divide that quotient by 3? Explain why or why not.

11. Rachel has 145 mugs displayed at a craft show. She displays them in 8 rows with the same number of mugs in each row. How many mugs are in each row? Explain how you interpreted the remainder. Use the same numbers and write a word problem in which the remainder is interpreted a different way.

12. **Test Practice** Valley Schools have a student population of 1,608 students. If there are an equal number of students in the 6 grade levels, how many students are there in each grade?

   Ⓐ 28 students   Ⓑ 208 students   Ⓒ 248 students   Ⓓ 268 students

Lesson 9
Place the First Digit

ESSENTIAL QUESTION
What strategies can be used to divide whole numbers?

A three-digit dividend may not have enough hundreds to divide. If so, the quotient should start at the next place-value position.

Math in My World

Example 1

Raven received 135 emails over 3 weeks. If she received the same number of emails each week, how many emails did she receive in the first week?

Find 135 ÷ 3.

Estimate 150 ÷ 3 = _____

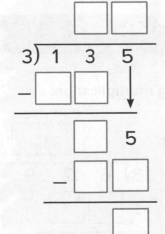

1 Divide the hundreds. There are not enough hundreds to divide into 3 groups. So, regroup 1 hundred and 3 tens as 13 tens.

2 Divide the tens. The first digit of the quotient is in the tens place.

3 Divide the ones.

So, Raven received _____ emails in the first week.

Check for Reasonableness Compare to the estimate.

_____ ≈ 50

Helpful Hint
The symbol ≈ *means about or almost equal to*.

Example 2

Find 7)6,784.

Estimate 7,000 ÷ 7 = _____

1 Divide the hundreds.
There are not enough
thousands to divide into
7 groups. So, regroup
6 thousands and
7 hundreds as
67 hundreds.

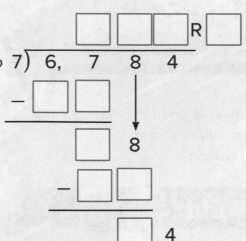

7)6, 7 8 4

8

4

2 Divide the tens.
The first digit of
the quotient is
in the hundreds
place.

3 Divide the ones.

So, 6,784 ÷ 7 = _____ .

Check for Reasonableness Compare to the estimate.

_____ ≈ _____

Talk MATH

You want to find 510 ÷ 6. Tell
how you know where to
place the quotient's first digit.

Guided Practice

Divide. Check your answer using multiplication.

1.

5)4 3 5

5

2.

8)6 2 9   R

9

Independent Practice

Divide.

3. $6\overline{)576}$

4. $5\overline{)3,085}$

5. $4\overline{)256}$

6. $6\overline{)4,527}$

7. $4\overline{)217}$

8. $4\overline{)274}$

9. $2,181 \div 3 =$ _____

10. $108 \div 9 =$ _____

11. $3,417 \div 4 =$ _____

Algebra Find the unknown number in each equation.

12. $232 \div 8 = q$

13. $324 \div 9 = s$

14. $192 \div 4 = y$

$q =$ _____

$s =$ _____

$y =$ _____

Problem Solving

15. There are 624 envelopes to be sorted into 8 different mail bags. If the same number of envelopes will be in each bag, how many envelopes will be in one bag?

_____

16. **Processes &Practices**  **2** **Use Symbols** There are 594 people standing in line to see a movie premiere. The movie is playing in 6 theaters. If the same number of people will see the movie in each theater, how many people will be in each theater? Write an equation to find the unknown. Then find the unknown.

_____

Brain Builders

17. The Environmental Club is having a trash pickup day. There are 130 people signed up to help. For the trash pickup day, they will work in groups of 4 people. No more than 4 people can join a group. How many groups are there? Explain how you interpreted the remainder.

_____

18. **Processes &Practices** **2** **Use Number Sense** Can you determine the number of digits in the quotient of 637 ÷ 7 without dividing? Explain how you know.

_____

_____

19. **Building on the Essential Question** How can I know where to place the first digit of a quotient? Write a division expression and explain each step to a friend.

_____

_____

Name ....

MY Homework

Homework Helper

Need help? ⟋ **connectED.mcgraw-hill.com**

Find 498 ÷ 6.

Estimate 500 ÷ 5 = 100

1 Divide the hundreds.
There are not enough hundreds to divide into 6 groups. So, regroup 4 hundreds and 9 tens as 49 tens.

```
        83
    6)498
     − 48↓
        18
      − 18
         0
```

2 Divide the tens.
The first digit of the quotient is in the tens place.

3 Divide the ones.

So, 498 ÷ 6 = 83.

Check Multiply to check your answer. 83 × 6 = 498

Practice

Divide.

1. 6)486

2. 6)392

3. 5,920 ÷ 6 = _____

Problem Solving

4. The phone company needs 420 poles to repair the telephone lines. Each truck holds 6 poles. How many trucks will they need?

_____

5. The festival committee has $1,544 to spend on pies for the pie-eating contest. If each pie costs $8, how many pies can the committee purchase?

_____

Brain Builders

6. Algebra A group of 485 people takes a canoe trip. They fill up all available canoes that each hold 3 people with 273 people. The rest of the group will ride in canoes that each hold 2 people. How many canoes will the group need? Write equations with unknowns to show your steps.

_____

_____

7. Processes & Practices **3** **Which One Doesn't Belong?** Circle the expression that does not have a two-digit quotient.

| $519 \div 6$ | $915 \div 7$ | $439 \div 7$ | $812 \div 9$ |

Explain how to use mental math to find the answer.

_____

8. Test Practice A glass company ships 470 glass ornaments. Each box holds 5 ornaments. How many boxes will the company need?

- Ⓐ 84 boxes
- Ⓑ 92 boxes
- Ⓒ 93 boxes
- Ⓓ 94 boxes

Lesson 10
Quotients with Zeros

ESSENTIAL QUESTION
What strategies can be used to divide whole numbers?

 Math in My World Watch | Tutor

Example 1

Maya is saving to buy a television. The television costs $327. She plans to save money for 3 months. How much does Maya need to save each month to buy the television?

Let m represent the amount Maya needs to save each month. Find the unknown in the equation $\$327 \div 3 = m$.

Estimate $\$300 \div 3 =$ _____

1 Divide the hundreds.

2 Divide the tens.

There are not enough tens to divide.

Place a 0 in the quotient.

3 Regroup the two tens as twenty ones.

There are now 27 ones.

4 Divide the ones.

So, $\$327 \div 3 =$ _____. Since $m =$ _____, Maya

needs to save _____ each month.

Check for Reasonableness Compare to the estimate. _____ ≈ _____

Example 2

Find 5,231 ÷ 4.

1 Divide the thousands.

2 Divide the hundreds.

3 Divide the tens.
There are not enough tens to divide. Place a 0 in the tens place.

4 Divide the ones.

The remainder is _____.

So, 5,231 ÷ 4 = _____.

Check Use multiplication to check your answer.

_____ × 4 = _____ and 5,228 + _____ = 5,231

$$\begin{array}{r} \boxed{},\boxed{}\,\boxed{}\,\boxed{}\ R\ \boxed{} \\ 4\overline{)5,\quad 2\quad 3\quad 1} \\ -\boxed{} \\ \hline \boxed{}\quad 2 \\ -\boxed{}\ \boxed{} \\ \hline \boxed{}\quad 3 \\ -\boxed{} \\ \hline \boxed{}\quad 1 \\ -\boxed{}\ \boxed{} \\ \hline \boxed{} \end{array}$$

Guided Practice

1. Divide.

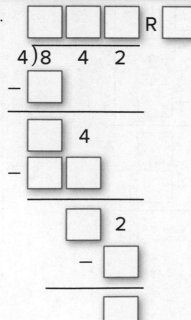

$$\begin{array}{r} \boxed{}\,\boxed{}\,\boxed{}\ R\ \boxed{} \\ 4\overline{)8\quad 4\quad 2} \\ -\boxed{} \\ \hline \boxed{}\quad 4 \\ -\boxed{}\ \boxed{} \\ \hline \boxed{}\quad 2 \\ -\boxed{} \\ \hline \boxed{} \end{array}$$

Talk MATH

Yolanda wants to find 936 ÷ 9. In which place-value position should she place a zero? Explain.

Independent Practice

Divide.

2. $2\overline{)418}$

3. $2\overline{)210}$

4. $4\overline{)4,324}$

5. $6\overline{)782}$

6. $2\overline{)6,213}$

7. $3\overline{)6,192}$

8. $840 \div 7 =$ _____

9. $627 \div 3 =$ _____

10. $5,330 \div 5 =$ _____

11. $8,017 \div 9 =$ _____

12. $413 \div 4 =$ _____

13. $9,163 \div 3 =$ _____

Problem Solving

14. **Processes &Practices** **2** **Use Algebra** There are 312 fish at the aquarium in 3 different fish tanks. Each tank has the same number of fish. How many fish are in each tank? Write an equation to find the unknown. Then find the unknown.

_____

15. There are 1,620 minutes of music to be put on 9 CDs. If the same number of minutes fits on each CD, how many minutes of music fit on each CD?

_____

Brain Builders

16. Gina spent 120 minutes helping her neighbors rake leaves and 240 minutes pulling weeds in the last 4 days. She helped the same amount of minutes each day. How many minutes did she work each day?

_____

17. **Processes &Practices** **1** **Keep Trying** Write two division problems that have zeros in the quotient. One of the problems should have a remainder and the other should not. Work each problem and explain each step to a friend.

_____

_____

_____

18. **Building on the Essential Question** How can I know when to place a zero in the quotient? Write and work a division expression to model the answer.

_____

_____

_____

Name _____

MY Homework

Homework Helper

eHelp

Need help? connectED.mcgraw-hill.com

Find 815 ÷ 2.

1 Divide the hundreds.

2 Divide the tens.
There are not enough tens to divide.
Place a 0 in the tens place.

3 Divide the ones.

The remainder is 1.

$$
\begin{array}{r}
407R1 \\
2\overline{)815} \\
-8 \\
\hline
01 \\
-0 \\
\hline
15 \\
-14 \\
\hline
1
\end{array}
$$

So, 815 ÷ 2 = 407 R1.

Check 407 × 2 = 814 and 814 + 1 = 815

Practice

Divide.

1. 8)856

2. 3)2,926

3. 841 ÷ 4 = _____

Problem Solving

4. Monica wants to join the swim team. She practices 812 minutes in 4 weeks. She practices the same number of minutes each week. How many minutes does Monica practice each week?

Brain Builders

5. Algebra The art teacher asks her students to cut out 1,045 apples and 1,463 leaves from construction paper. Five apples can be cut from a sheet of paper. Seven leaves can be cut from a sheet of paper. How many sheets of paper does she need? Write equations to find the unknown. Then find the unknown.

6. **Processes &Practices** **3** **Which One Doesn't Belong?** Circle the division problem that does not belong with the other three. Explain. Write a problem that does belong.

621 ÷ 6	384 ÷ 3	719 ÷ 7	514 ÷ 5

7. Test Practice Abby and her family are going to Yellowstone National Park this summer. They drive 1,212 miles from their home to the park. If they drive the same number of miles each day for 4 days, how many miles will they drive each day?

Ⓐ 303 miles 　Ⓒ 403 miles

Ⓑ 330 miles 　Ⓓ 3,030 miles

Lesson 11
Hands On

Use Models to Interpret the Remainder

ESSENTIAL QUESTION
What strategies can be used to divide whole numbers?

Build It Tools

A group of fifth graders collected 46 cans of food to donate to 3 food banks. If each food bank is to get an equal number of cans, how many cans do they each receive?

1 Use _____ connecting cubes to represent the cans of food.

Use _____ paper plates to represent the food banks.

Divide the cubes equally among the _____ plates.

How many cubes are on each plate? _____

How many cubes are left over? _____

2 Interpret the remainder.
Since each food bank is to get the same number of cans of food, they will each receive

_____ cans.

There is _____ can left over.

A total of 35 students are going on a field trip to NASA's Johnson Space Center in Houston, Texas. If there needs to be one adult for every 8 students, how many adults are needed?

 Use _____ connecting cubes to represent the students.

Use paper plates to represent the adults.

How many plates are there with 8 students? _____

How many cubes are left over? _____

 Interpret the remainder.

There are _____ groups of _____ students. They will each need an adult.

There are _____ students who are not enough for a full group of 8. They will also need an adult.

So, _____ + _____ , or _____ adults are needed.

Talk About It

1. **Processes &Practices** 🔢 **Justify Conclusions** In Activity 1, the remainder was dropped. Explain why.

_____

_____

_____

2. In Activity 2, the quotient was "rounded up" to 5. Explain why.

_____

_____

Practice It

Solve using models. Explain how to interpret the remainder. Draw your models.

3. Each picnic table at a park seats 6 people. How many tables will 83 people at a family reunion need?

_____

_____

4. Mrs. Malone has $75 to buy volleyballs for Lincoln Middle School. How many can she buy at $9 each?

_____

_____

5. Darcy has 63 flowers to use to make centerpieces. Each centerpiece uses 8 flowers. How many centerpieces can she make?

_____

_____

Apply It

For Exercises 6–8, solve using models. Explain how to interpret the remainder.

6. A teacher received 30 new calculators. Only 8 calculators fit in each carrier. How many carriers does the teacher need?

_____

_____

7. **Processes & Practices** ➡ **5** **Use Math Tools** Joel has 48 oranges. He puts 7 oranges in a bag. How many bags can he fill?

_____

_____

8. Anita is helping to make gift boxes for a community center. She has a total of 34 toys. She places 3 toys in each box. How many boxes will she need?

_____

_____

9. **Processes & Practices** ➡ **2** **Reason** Suppose 2 friends want to share 5 cookies evenly. Interpret the remainder in two different ways.

_____

_____

Write About It

10. How can I use models to interpret the remainder?

_____

_____

Name ....

Homework Helper

Need help? connectED.mcgraw-hill.com

A total of 47 students signed up to play soccer. If there are 4 teams, how many students are on each team?

1 Use 47 connecting cubes to represent the students. Use 4 paper plates to represent the teams. How many cubes are on each plate? 11
How many cubes are left over? 3

2 Interpret the remainder.
There are 4 teams of 11 students. There are 3 students who are left over and will be placed on teams.

So, 3 teams will have 12 students and 1 team will have 11 students.

Practice

1. Jarrod's sports card holder can hold 9 cards on each page. How many pages will Jarrod need if he has 75 sports cards? Solve using models. Explain how to interpret the remainder. Draw your models.

   _____

   _____

Solve using models. Explain how to interpret the remainder. Draw your models.

2. A group of 4 students are selling candy bars to raise money for a field trip. The group needs to sell 53 candy bars. If each student sells an equal amount, how many candy bars do they each sell? Solve using models. Explain how to interpret the remainder. Draw your models.

_____

_____

Problem Solving

For Exercises 3–5, solve using models. Explain how you interpreted the remainder. Draw your models.

3. **Processes &Practices** **4** **Model Math** Joel has 59 songs on his MP3 player. He equally divides them in 7 groups. How many songs will be in each group?

_____

_____

4. Lauren's shoe organizer can hold 5 pairs of shoes in each row. How many rows will Lauren need if she has 24 pairs of shoes?

_____

_____

5. Mr. Staley has $62 to buy binders for his mathematics class. How many binders can he buy at $3 each?

_____

_____

Lesson 12
Interpret the Remainder

 Math in My World Watch Tutor

Example 1

A state park has **257** evergreens to plant equally in 9 areas. How many evergreens are planted in each area? What does the remainder represent?

Divide 257 ÷ 9.

1 Place the first digit.

$25 \div 9 \approx 2$

Put 2 in the tens place of the quotient.

2 Multiply. $9 \times 2 = 18$

Subtract. $25 - 18 = 7$

Compare. $7 < 9$

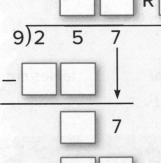

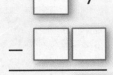

3 Divide the ones.

$77 \div 9 \approx 8$

$9 \times 8 = 72$

$77 - 72 = 5$

$5 < 9$

4 Write the remainder.

The remainder is 5.

5 Interpret the remainder, 5.

The remainder, 5, means there are 5 evergreens left over.

So, the park plants _____ evergreens in each area and _____ evergreens are left.

Example 2

There are 174 guests invited to a dinner. Each table seats 8 guests. How many tables are needed?

Divide 174 ÷ 8.

 Place the first digit.

17 ÷ 8 ≈ 2

Put 2 in the tens place of the quotient.

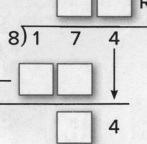
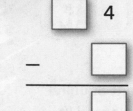

 Multiply. 8 × 2 = 16

Subtract. 17 − 16 = 1

Compare. 1 < 8

 Divide the ones.

14 ÷ 8 ≈ 1

1 × 8 = 8

14 − 8 = 6

6 < 8

4 Write the remainder.

The remainder is 6.

 Interpret the remainder, 6.

There are 6 guests left over, which is not enough for a full table of 8. But, they also need a table.

Talk MATH

Discuss the different ways you can interpret the remainder.

So, a total of _____ + _____ , or _____ tables are needed.

Guided Practice

1. A tent is put up with 7 poles. How many tents can be put up with 200 poles?

Divide 200 ÷ 7.

Interpret the remainder, _____.

There are _____ poles left over, which is not enough for another tent.

So, a total of _____ tents can be put up.

Independent Practice

Solve. Explain how you interpreted the remainder.

2. There are 50 students traveling in vans on a field trip.
Each van seats 8 students. How many vans are needed?

_____

_____

3. How many payments of $10 would it take Samuel
to purchase the scooter shown at the right?

_____

_____

4. Mrs. Hodges made 144 muffins for a bake sale. She puts them into
tins of 5 muffins each. How many tins of muffins can she make?

_____

_____

5. How many 8-foot sections of fencing are needed for 189 feet
of fence?

_____

_____

Problem Solving

Solve Exercises 6–8. Explain how you interpreted the remainder.

6. **Processes &Practices** **2** **Reason** Students on the softball team earned $295 from a car wash. How many team banners can they buy if each banner costs $8?

7. Valerie has 20 stuffed animals. She wants to store them in plastic bags. She estimates she can fit three stuffed animals in each bag. How many bags will she need?

8. How many 6-ounce cups can be filled from 4 gallons of juice? (*Hint:* 1 gallon = 128 ounces)

Brain Builders

9. **Processes &Practices** **1** **Make Sense of Problems** If the divisor is 30, what is the least three-digit dividend that would give a remainder of 8? Explain.

10. **Building on the Essential Question** How can I interpret the remainder? Give two examples of interpreting the remainder in different ways.

MY Homework

Homework Helper

Need help? connectED.mcgraw-hill.com

There are 46 students volunteering at a senior citizen community. There is a maximum of 6 students in each group. How many groups are needed?

Divide 46 ÷ 6.

1 Place the first digit.

$46 \div 6 \approx 7$
Put 7 in the ones digit of the quotient.

$$
\begin{array}{r}
7\,R4 \\
6)\overline{46} \\
-\,42 \\
\hline
4
\end{array}
$$

2 Multiply. $6 \times 7 = 42$
Subtract. $46 - 42 = 4$
Compare. $4 < 6$

3 Write the remainder.

The remainder is 4.

4 Interpret the remainder, 4.

There are 4 students left over, which is not enough for another group of 6. But, they are volunteering.

So, there are a total of 7 + 1, or 8 groups of students volunteering.

Practice

1. Four employees of Papa Tony's Pizza are cleaning up at the end of a busy night. There is a list of 43 clean-up tasks that need to be completed. If each employee does the same number of tasks, how many tasks should each employee do? Solve. Explain how you interpreted the remainder.

_____

_____

Problem Solving

Solve Exercises 2 and 3. Explain how you interpreted the remainder.

2. Mrs. Flores is buying scrapbooks for her store. Her budget is $350. Each scrapbook costs $9. How many scrapbooks can she buy?

_____

_____

Brain Builders

3. Water stations will be placed every 400 meters of a 5-kilometer race. How many water stations are needed? (*Hint:* 1 kilometer = 1,000 meters) Draw a diagram to model the problem and explain your diagram to a friend.

_____

_____

_____

4. **Processes &Practices** 4 **Model Math** Write a real-world problem that is represented by the division problem 38 ÷ 5 = 7 R3 in which it makes sense to round the quotient up to 8.

_____

_____

5. **Test Practice** Three yards of fabric will be cut into pieces so that each piece is 8 inches long. How many pieces can be cut?
(*Hint:* 1 yard = 36 inches)

Ⓐ 4 pieces with 4 inches left over

Ⓑ 10 pieces with 6 inches left over

Ⓒ 13 pieces with 4 inches left over

Ⓓ 16 pieces with 6 inches left over

Lesson 13
Problem-Solving Investigation
STRATEGY: Determine Extra or Missing Information

ESSENTIAL QUESTION
What strategies can be used to divide whole numbers?

Learn the Strategy

Watch Tutor

Kayla was collecting book orders. The cost of each book is $3. There were 7 orders on Wednesday, 5 orders on Thursday, and more orders on Friday and Monday. How many book orders were collected altogether?

1 Understand

What facts do you know?

I know the cost of a book is $3 and the number of book orders

on Wednesday was _____ and Thursday was _____.

What do you need to find?

I need to find the total number of _____.

2 Plan

Determine if there is extra or missing information.

The _____ of a book is not needed. The number of book orders

that were collected on Friday and Monday is missing.

3 Solve

Some information is _____. So, I cannot solve the problem.

4 Check

Is my answer reasonable? Explain.

_____

Rocco is slicing a loaf of Italian bread for dinner. The bread costs $4. He cuts the loaf into slices that are 1 inch thick. If the loaf is 18 inches long, how many pieces of bread did he cut?

 **Understand**

What facts do you know?

_____

_____

_____

What do you need to find?

_____

 **Plan**

_____

_____

_____

Solve

Check

Is my answer reasonable? Explain.

_____

_____

Name ....

Apply the Strategy

Determine if there is extra or missing information. Then solve the problem, if possible.

1. Jayden is downloading songs onto his MP3 player. One song is 5 minutes long, another is 2 minutes long, and a third is between the lengths of the other two songs. What is the total length of all three songs?

   _____

2. Room 220 and Room 222 are having a canned food drive. How many more cans has Room 222 collected than Room 220?

   _____

ROOM 222 346 CANS

ROOM 220 278 cans

Brain Builders

3. Karly is collecting money for a bowl-a-thon. Her goal is to collect $125. Last year the bowl-a-thon raised $100. If she charges $5 for each person, how many people need to participate in the bowl-a-thon?

   _____

4. **Processes &Practices** **1** **Make Sense of Problems** Sari made pancake batter. She has 1 cup of batter left. How much batter did she use? If this problem cannot be solved with the information provided, write additional information and then solve the problem.

   _____

   _____

   _____

Review the Strategies

Use any strategy to solve each problem.

- Determine extra or missing information.
- Make a table.
- Use the four-step plan.

5. Mrs. Rollins raises prize chickens. Each chicken eats the same amount of food. Mrs. Rollins bought 100 pounds of chicken food last week. How much food did each chicken eat?

6. At the school bake sale, Kenji's mom bought 3 cookies, 1 brownie, and 1 cupcake. She gave the cashier $2 and received $1.05 in change. Find the cost of a cupcake and write it in the table.

Item	Price ($)
cookie	0.15
brownie	0.20
cupcake	

Processes
7. &Practices **1** **Plan Your Solution** Todd has $50 to spend on a video game. The game he wants costs $30. If he buys one game, he gets a second game for half price. How much money will he have left if he purchases two games?

8. Twelve students are going roller skating. Each student pays $8 for a ticket and $4 for snacks. Find the total cost for tickets and snacks.

9. The table shows the number of miles the Wong family drove each day on their vacation. How many more miles did they drive on day 1 than on day 4?

Day	Miles
1	345
2	50
3	89
4	279

MY Homework

Homework Helper

Need help? connectED.mcgraw-hill.com

Anna collected 50 cans for a food drive. She collected
10 cans each day of the drive, mostly green beans.
How many days did she collect cans?

1 Understand

What facts do you know?

- I know that Anna collected 50 cans.

- I know that each day of the drive she collected 10 cans.

What do you need to find?

- I need to find the number of days she collected cans.

2 Plan

Determine if there is extra or missing information.

The information that Anna collected mostly green beans is not needed.

I have all the necessary information to solve the problem.

3 Solve

$50 \div 10 = 5$ days

So, Anna collected cans for 5 days.

4 Check

Is my answer reasonable? Explain.

$5 \times 10 = 50$ cans

My answer is reasonable.

Practice

Determine if there is extra or missing information. Then solve the problem, if possible.

1. Mrs. Blackwell gives each of her students two pencils. How many pencils did she hand out?

_____

2. If David plays 3 tennis matches every week for 9 weeks, how many matches will he play altogether?

_____

3. The Alvarez family bought a car for $2,000. They made a down payment of $500. If they want to pay the balance in 5 equal payments, how much will each of these payments be?

_____

Brain Builders

4. **Processes &Practices** ➊ **Make Sense of Problems** Melanie runs 5 miles a day. How many miles will her brother run in one week? If this problem is missing information, rewrite the problem and solve.

_____

_____

5. Marco does 14 extra math problems each school night. How many extra problems does he do each week? There are 5 school nights each week. What is the difference in the number of extra math problems Marco works each night if he works the same total number, but he does it over the course of 7 days? Explain.

_____

_____

Vocabulary Check

Complete the crossword puzzle using the words in the word bank.

dividend   divisor   fact family

partial quotients   quotient   remainder

Across

1. the number that divides the dividend

2. a method of dividing where you break the dividend into sections that are easy to divide

3. the result of a division problem

4. a number that is being divided

Down

5. the number that is left after one whole number is divided by another

6. a group of related facts using the same numbers

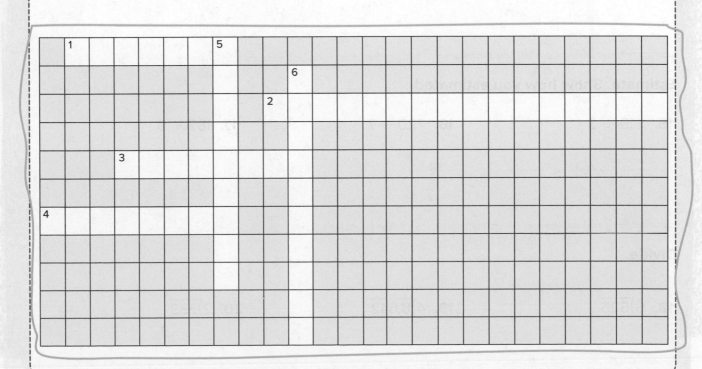

Concept Check

Divide. Use a related multiplication fact.

7. $14 \div 7 =$ _____

8. $40 \div 5 =$ _____

Find the quotient using a model.

9. $48 \div 3 =$ _____

Divide.

10. $3\overline{)93}$

11. $5\overline{)78}$

12. $2\overline{)47}$

Divide mentally.

13. $300 \div 3 =$ _____

14. $160 \div 8 =$ _____

Estimate. Show how you estimated.

15. $219 \div 2$

16. $720 \div 7$

17. $182 \div 8$

Divide.

18. $5\overline{)625}$

19. $4\overline{)8,642}$

20. $2\overline{)533}$

Problem Solving

21. Melanie swam 918 meters in 3 days. If she swam the same distance each day, how far did Melanie swim in one day?

_____

22. Luke has 48 books. He puts 7 books in a box. How many boxes can he fill? Explain how you interpreted the remainder.

_____

_____

23. In 2 hours Cara read 48 pages. If she read the same number of pages each hour, how many pages did Cara read in one hour?

_____

Brain Builders

24. Three plane tickets to New York cost $2,472. If each plane ticket costs the same amount, about how much does one ticket cost? How much would the airline have to discount each ticket for the new total to be $2,400? Show your work.

_____

_____

25. Carmen has 468 trading cards in 4 binders. If each binder has the same number of cards, how many cards are in each binder? If 9 cards fit on a page, how many pages are in all 4 binders?

_____

_____

26. Test Practice A cabinet with 4 shelves can hold 1,640 CDs. If the shelves each hold the same number of CDs, how many CDs does each shelf hold?

Ⓐ 400 CDs   Ⓒ 420 CDs

Ⓑ 410 CDs   Ⓓ 424 CDs

Use what you learned about dividing by a one-digit divisor to complete the graphic organizer.

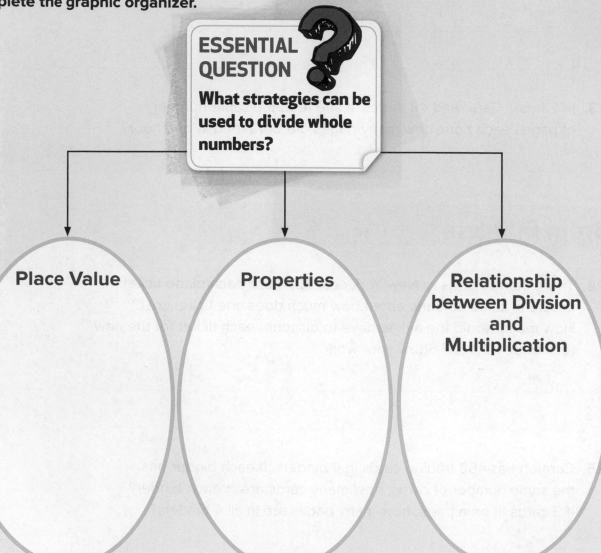

ESSENTIAL QUESTION

What strategies can be used to divide whole numbers?

Place Value

Properties

Relationship between Division and Multiplication

Now reflect on the ESSENTIAL QUESTION Write your answer below.

_____

_____

_____

Performance Task

Brain Builders

Saving for a Field Trip

Mrs. Bell's class is working together to earn enough money for a field trip. The total cost for the class to go to an observatory is $910. The table shows the three different ways they will earn the money.

Raking and Packing Leaves for Pick Up	$7 per bag
Make Tie Dye T-Shirts	$8 profit per shirt
Helping an Adult	$6 per 30 minutes

Show all your work to receive full credit.

Part A

The students want to have the money for the field trip in 5 months. How much do they need to earn each month in order to accomplish their goal? Explain.

Part B

In the first month, the students only rake and pack leaves for pick up. How many bags do the students need to fill in order to make their goal for the first month? Explain.

Part C

In the second and third months, the students only make tie dye T-shirts to sell. How many T-shirts do they need to sell in order to make their goal? Explain the meaning of the remainder.

Part D

How much do the students have left to earn? They only earn money as helpers for 30 minutes in the last two months. How many times will they have to help an adult for 30 minutes in order to make their final goal? Explain.

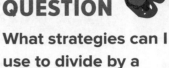

Around My School

Watch a video!

Brain Builders

MY Chapter Project

Plan a Field Trip!

1. Brainstorm with your group different places for the class to go on a field trip. List the places in the space below.

2. As a group, select one of the places and write it below.

Field Trip location: _____

3. Create a 2-column table in the space below. List all the costs for the trip, such as transportation, food, and other expenses, in the left column of the table. Use the Internet or other resources to find the amount of each cost for the trip. List the amounts, in dollars, in the right column of the table.

4. Divide the total cost of the field trip by the number of students in your class to find how much money each student will pay.

Name _____

Am I Ready?

Estimate each product. Tell whether the estimate is *greater than* or *less than* the actual product.

1. $224 × 12 = _____

2. 372 × 36 = _____

3. 488 × 85 = _____

4. 515 × 41 = _____

Multiply.

5. 14 × 3 = _____

6. 36 × 5 = _____

7. 76 × 4 = _____

8. The teacher purchased 13 packs of crayons. There are 24 crayons in each pack. How many crayons are there in all?

_____

9. A movie was sold out for four straight days. If 535 tickets were sold each day, how many tickets were sold in all?

_____

Shade the boxes to show the problems you answered correctly.

How Did I Do? | 1 | 2 | 3 | 4 | 5 | 6 | 7 | 8 | 9 |

MY Math Words

Vocab
abc

Review Vocabulary

| dividend | divisor | quotient |

Making Connections

Use the review vocabulary to label the examples in the second and third columns. Then write 2 more real-world problems.

Real-World Problem	Show ÷	Show ⟌
Thirty-two students were asked to make 8 signs for a car wash. How many groups did the students work in if each group made 1 sign and had an equal number of students?	32 ÷ 8 = 4	$\frac{4}{8\smash{)}32}$
	48 ÷ 6 = 8	$\frac{8}{6\smash{)}48}$
	45 ÷ 15 = 3	$\frac{3}{15\smash{)}45}$

MY Vocabulary Cards

Processes & Practices

Ideas for Use

- Use the blank cards to write review vocabulary terms that relate to division of whole numbers.

- Write the name of each lesson on the front of each card. On the back, include a few study tips or important concepts to remember.

MY Foldable

FOLDABLES® Follow the steps on the back to make your Foldable.

✂

Write the problem	Estimate	Divide	Check

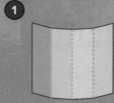

Divide

Estimate

Write
the problem

Lesson 1
Estimate Quotients

ESSENTIAL QUESTION
What strategies can I use to divide by a two-digit divisor?

You can use rounding and compatible numbers to estimate quotients when dividing by two-digit divisors. By estimating first, you can determine the reasonableness of your results.

 Math in My World Watch Tutor

Example 1

The school principal has 812 fliers to pass out equally to 19 different teachers. About how many fliers would each teacher receive?

Estimate 812 ÷ 19.

1 Round the divisor to the nearest ten.

$$812 \div 19$$

2 Round the dividend to the nearest hundred.

$$812 \div \underline{\hspace{2cm}}$$

$$\underline{\hspace{2cm}} \div \underline{\hspace{2cm}}$$

3 Divide mentally.

$$\underline{\hspace{1cm}} \div \underline{\hspace{1cm}} = \underline{\hspace{1cm}}$$

$$8 \div 2 = 4$$
$$80 \div 20 = 4$$
$$800 \div 20 = 40$$

So, each teacher would receive about _____ fliers.

Online Content at <a>connectED.mcgraw-hill.com</a>

Example 2
Tutor

Estimate 234 ÷ 41.

 Round the divisor to the nearest ten.

41 ⟶ _____

Helpful Hint
Another way to write 234 ÷ 41 is
41)‾2‾3‾4‾

 Change the dividend to a number that is compatible

with the rounded divisor, _____ .

234 ⟶ _____

 Divide mentally.

_____ ÷ _____ = _____

So, 234 ÷ 41 is about _____ .

Check Use multiplication to check your answer.

_____ × _____ = _____ and _____ ≈ 234

Guided Practice

1. Estimate 312 ÷ 31. Show how you estimated.

   Round the divisor to the nearest ten.

   31 ⟶ _____

   Round the dividend to the nearest hundred.

   312 ⟶ _____

   Divide mentally.

   _____ ÷ _____ = _____

   So, 312 ÷ 31 is about _____ .

Talk MATH

Is it possible to have more than one estimate for a division problem? Explain. Give an example.

Independent Practice

Estimate by rounding. Show how you estimated.

2. $121 \div 42$

3. $400 \div 23$

4. $642 \div 83$

5. $28\overline{)597}$

6. $38\overline{)244}$

7. $24\overline{)943}$

Estimate using compatible numbers. Show how you estimated.

8. $653 \div 52$

9. $208 \div 51$

10. $300 \div 59$

11. $32\overline{)619}$

12. $43\overline{)847}$

13. $34\overline{)272}$

Problem Solving

14. There are 598 goldfish divided equally among 23 fish tanks. About how many goldfish are in each tank?

_____

15. The area of a rectangle is 138 square meters, and the length is 21 meters. About how many meters long is the width?

_____

Brain Builders

16. **Processes &Practices** **Check for Reasonableness** A box of cereal contains 340 grams of carbohydrates. If there are 12 servings, about how many grams are there in one serving? Show two ways to estimate. Explain why your estimates are reasonable.

_____

_____

17. **Processes &Practices** **3** **Which One Doesn't Belong?** Circle the equation that is not a reasonable estimate for 533 ÷ 57. How can you know when an estimate is not reasonable?

$$540 \div 60 = 9 \qquad 500 \div 50 = 10$$

$$550 \div 55 = 10 \qquad 420 \div 60 = 7$$

_____

_____

18. **?** **Building on the Essential Question** Explain when it would be useful to estimate. Give an example.

_____

_____

Name ....

MY Homework

Lesson 1

Estimate Quotients

Homework Helper

Need help? 🖝 **connectED.mcgraw-hill.com**

Estimate 304 ÷ 18.

1 Round the divisor to the nearest ten. $18 \longrightarrow 20$

2 Change the dividend to a number that is $304 \longrightarrow 300$
compatible with the rounded divisor, 20.

3 Divide mentally. $300 \div 20 = 15$

So, 304 ÷ 18 is about 15.

Check Use multiplication to check your answer.

$15 \times 20 = 300$ and $300 \approx 304$

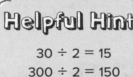

Helpful Hint

$30 \div 2 = 15$
$300 \div 2 = 150$
$300 \div 20 = 15$

Practice

Estimate. Show how you estimated.

1. 512 ÷ 52   **2.** 412 ÷ 97

3. 83)237   **4.** 31)458

Copyright © McGraw-Hill Education

Lesson 1 My Homework   **255**

![Real World] **Problem Solving**

5. The Booster Club is having a bake sale. The members place 25 baked goods in each bag. There are 630 baked goods donated. About how many bags will the club have for sale?

_____

6. A farmer has 212 acres of land for sale. He divides the land into 18 equal sections. About how many acres are in each section?

_____

7. Alexander has 418 songs on his MP3 player. He divides the songs into 11 equal groups. About how many songs are in each group?

_____

Brain Builders

8. **Processes &Practices** **1** **Check for Reasonableness** A restaurant ordered 833 ounces of chicken. There are 16 ounces in a pound. About how many pounds of chicken did the restaurant order? Explain why your estimate is reasonable.

_____

_____

9. Test Practice Ms. Robbins has 615 sheets of paper for a project. She has 48 students. Which is the best estimate for the number of sheets of paper she can give to each student?

Ⓐ 10 sheets   Ⓒ 15 sheets

Ⓑ 12 sheets   Ⓓ 20 sheets

Lesson 2
Hands On
Divide Using Base-Ten Blocks

ESSENTIAL QUESTION
What strategies can I use to divide by a two-digit divisor?

Build It

Gary is saving up to buy a trumpet for band that costs $156. Suppose he saves the same amount each month for 12 months. How much money does he need to save each month?

Find 156 ÷ 12. Use base-ten blocks to find the quotient.

1 Model 156 using base-ten blocks.

2 Since you can not separate the hundreds block into 12 groups, regroup it into tens.

There are _____ tens total.

3 Divide the tens equally into 12 groups. Circle each group.

How many tens are in each group? _____

There are still _____ tens and _____ ones that need to be divided.

4 Use the remaining tens and ones and regroup them as ones. Then divide the ones equally into 12 groups. Draw the results in the white space below.

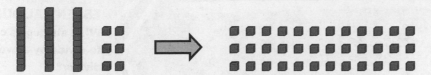

There are _____ ones.

How many ones are in each group? _____

Each group contains _____ ten and

_____ ones, or _____ .

So, Gary needs to save _____ each month.

Check Use multiplication to check your answer.

_____ × 12 = $156

Talk About It

1. In the activity, you started by placing 1 ten in each group. What will be the place value of the first digit of the quotient?

_____

2. What would happen if the cost of the trumpet was $168? Would the amount to save each month increase or decrease?

_____

3. **Processes &Practices** **3** **Justify Conclusions** Suppose Gary chose to save $156 for 13 months instead of 12 months. Will the amount he needs to save per month increase or decrease? Explain your answer.

_____

_____

_____

Practice It

Use models to find each quotient. Draw the equal groups.

4. $117 \div 13 =$ _____

5. $136 \div 17 =$ _____

6. $231 \div 11 =$ _____

7. $105 \div 15 =$ _____

Apply It

8. The average person eats 26 pounds of bananas each year. How many years would it take a person to eat 104 pounds of bananas? Draw models to find the quotient.

9. A travel van holds 11 people. There are a total of 143 people signed up to take a trip to the zoo. How many vans are needed? Draw models to find the quotient.

10. **Processes &Practices** 1 **Plan Your Solution** Cheryl has a collection of sports cards and has 8 pages in the album. Each page of the album holds 14 sports cards. If she completely fills the album, how many total sports cards does she have? Draw models to find the dividend.

11. **Processes &Practices** 5 **Use Math Tools** The average person eats 16 pounds of apples each year. How many years would it take a person to eat 144 pounds of apples? Draw models to find the quotient.

Write About It

12. How can base-ten blocks be used to divide by a two-digit divisor? Explain.

MY Homework

Homework Helper

Need help? ⟋ **connectED.mcgraw-hill.com**

Joey has 120 stamps. He puts an equal number of stamps on each of the 10 pages in an album. How many stamps are on each page?

Find 120 ÷ 10. Use base-ten blocks to find the quotient.

1 Base-ten blocks are used to model 120.

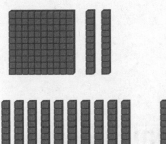

2 Since you can not separate the hundreds block into 10 groups, it was regrouped Into tens. There are 12 tens total.

3 The tens were divided equally into 10 groups shown by the circles.

There is 1 ten in each group.

There are 2 tens that still need to be divided.

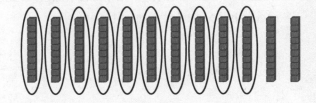

4 The remaining tens were regrouped as ones. They were divided equally into 10 groups.

There are 20 ones.

There are 2 ones in each group.

Each group contains 1 ten and 2 ones, or 12.

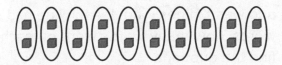

So, Joey can place 12 stamps on each page.

Check Use multiplication to check your answer. 12 × 10 = 120

Practice

Use models to find each quotient. Draw the equal groups.

1. $121 \div 11 =$ _____

2. $153 \div 17 =$ _____

Problem Solving

3. Braydon bought 13 packages of golf balls for $273. Each package cost the same amount. How much does one package cost? Draw models to find the quotient.

_____

4. Desirée wrote an essay for school that had a total of 247 words and was 13 lines long. If each line had an equal number of words, how many words were on each line? Draw models to find the quotient.

_____

Processes
5. &Practices **5** **Use Math Tools** There were 242 fish in 22 fish tanks at the pet store. If there were an equal number of fish in each fish tank, how many fish were in each tank? Draw models to find the quotient.

_____

Lesson 3
Divide by a Two-Digit Divisor

ESSENTIAL QUESTION
What strategies can I use to divide by a two-digit divisor?

 Math in My World Watch Tutor

Example 1

The yearbook committee took **836** photos with a digital camera over **76** days. If they took an equal amount of photos each day, how many photos did they take each day? Check your answer for reasonableness.

Let *p* represent the number of photos taken each day. Write an equation to find the value of *p*.

_____ ÷ _____ = *p*

Estimate $800 \div 80 =$ _____   So, the first digit is in the tens place.

1 Divide the tens.

$83 \div 76 \approx 1$

Write 1 in the quotient over the tens place.

2 Multiply. $76 \times 1 = 76$

Subtract. $83 - 76 = 7$

Compare. $7 < 76$

$76 \overline{)8 \quad 3 \quad 6}$

3 Bring down 6 ones.

There are 76 ones in all.

4 Divide the ones.

$76 \div 76 = 1$

Write 1 in the quotient over the ones place.

$76 \times 1 = 76$

$76 - 76 = 0$

So, $836 \div 76 =$ _____ . Since $p =$ _____ , the yearbook

committee will take _____ photos each day.

Check for Reasonableness _____ ≈ _____

Online Content at  **connectED.mcgraw-hill.com**

Example 2

Find 751 ÷ 30.

Estimate 750 ÷ 30 = _____

1 Divide the tens.

75 ÷ 30 ≈ 2

Write 2 in the quotient over the tens place.

2 Multiply. 30 × 2 = 60

Subtract. 75 − 60 = 15

Compare. 15 < 30

 R

$$30\overline{)751}$$

  1

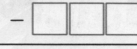

3 Bring down 1 one.

There are 151 ones in all.

4 Divide the ones.

151 ÷ 30 ≈ 5

Write 5 in the quotient over the ones place.

30 × 5 = 150

151 − 150 = 1

So, 751 ÷ 30 is _____ R _____.

Check for Reasonableness _____ ≈ _____ R _____

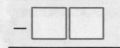

Explain how estimation is used to help you place the first digit in the quotient.

Guided Practice

1. Find 176 ÷ 16.   **Estimate**   180 ÷ 20 = _____

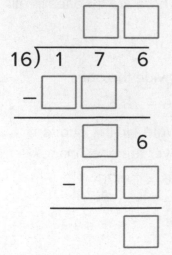

$$16\overline{)176}$$

So, 176 ÷ 16 = _____.

Check for Reasonableness _____ ≈ _____

Independent Practice

Divide. Check for reasonableness.

2. 809 ÷ 62 = _____

3. 925 ÷ 42 = _____

4. 210 ÷ 15 = _____

5. 27)‾873‾

6. 34)‾594‾

7. 12)‾155‾

8. 29)‾790‾

9. 18)‾416‾

10. 42)‾624‾

Algebra Divide to find the variable in each equation.

11. 840 ÷ 24 = h

h = _____

12. 528 ÷ 12 = b

b = _____

13. 952 ÷ 28 = w

w = _____

Problem Solving

14. Mr. Calzada buys flags for his store. Each flag costs $28. How many flags can he buy for $350? What does the remainder represent?

_____

15. The area of a rectangle is 384 square feet and the width is 24 feet. Find the length.

_____

16. Dreanne uploads 292 pictures to her online album. Her online album shows 12 pictures on each page. How many pages does she scroll through to see all 292 pictures? What does the remainder represent?

_____

Brain Builders

17. **Processes &Practices**  **Identify Structure** In 31 days, Gwen's dog sleeps 496 hours. If she sleeps the same number of hours each night, how many hours does she sleep per week? Write equations to solve, using h as the unknown. Explain how to use multiplication to check your answer.

_____

_____

18. **Processes &Practices** **1** **Make a Plan** Write a division problem with a quotient that is greater than 40 but less than 50.

_____

Write another division problem with a quotient that is greater than 20 but less than 25.

_____

19. **Building on the Essential Question** What is a standard procedure for dividing by a two-digit divisor? Explain.

_____

Homework Helper

Need help? connectED.mcgraw-hill.com

Find 204 ÷ 12.

Estimate 200 ÷ 10 = 20

1 Divide the tens.

20 ÷ 12 ≈ 1

Write 1 in the quotient
over the tens place.

$$
\begin{array}{r}
1\ 7 \\
12\overline{)2\ 0\ 4} \\
-\ 1\ 2 \\
\hline
8\ 4 \\
-\ 8\ 4 \\
\hline
0
\end{array}
$$

2 Multiply. 12 × 1 = 12
Subtract. 20 − 12 = 8
Compare. 8 < 12

3 Bring down 4 ones.
There are 84 ones
in all.

4 Divide the ones.

84 ÷ 12 = 7

Write 7 in the quotient
over the ones place.

12 × 7 = 84

84 − 84 = 0

So, 204 ÷ 12 is 17.

Check for Reasonableness 20 ≈ 17

Practice

Divide. Check for reasonableness.

1. 874 ÷ 23 = _____

2. 988 ÷ 96 = _____

3. 58)940

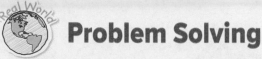

Problem Solving

4. **Algebra** Find the variable in the equation $803 \div 73 = m$.

   $m =$ _____

5. **Algebra** Find the variable in the equation $988 \div 76 = d$.

   $d =$ _____

Brain Builders

6. **Processes &Practices** **2** **Stop and Reflect** Members of the Bladerunners skating club collected $950 from fundraising activities. They want to buy Ultrablade skates, which are $48 a pair. How many pairs of skates can they buy? What does the remainder represent?

   _____

   _____

7. A theater has 990 total seats. There are r rows of seats in the theater. If each row has 45 seats, how many rows of seats are in the theater? Write a division equation with the unknown, then show the steps you use to solve.

   _____

8. A group of friends equally split the dinner bill shown. If each person paid $16, how many friends paid?

   _____

Restaurant

Chicken Tacos.......$82
Drinks...................$28
Salads...................$52
Desserts................$33
Tip.........................$29

9. **Test Practice** Emily loves to read. She read 527 hours in 31 weeks. If she read an equal number of hours each week, how many hours did she read each week?

   Ⓐ 10 hours   Ⓒ 15 hours

   Ⓑ 14 hours   Ⓓ 17 hours

Check My Progress

Vocabulary Check

Draw a line to match each vocabulary term with the correct definition.

1. **compatible numbers**

2. **remainder**

3. **dividend**

4. **estimate**

5. **quotient**

• a number close to an exact value which indicates about how much

• the result of a division problem

• the number that is left after one whole number is divided by another

• numbers in a problem that are easy to work with mentally

• a number that is being divided

Concept Check

Estimate. Show how you estimated.

6. $43\overline{)412}$

7. $81\overline{)637}$

8. $595 \div 28$

9. $22\overline{)311}$

Divide. Check for reasonableness.

10. $528 \div 16 =$ _____

11. $821 \div 22 =$ _____

12. $14\overline{)634}$

13. $18\overline{)954}$

Problem Solving

14. Algebra Find the variable in the equation $975 \div 39 = k$.

$k =$ _____

15. A movie theater has 576 seats arranged in 36 equal rows. How many seats are in each row?

_____

Brain Builders

16. The Mitchell family is making payments on the refrigerator to the right. If they pay $41 every month for 2 years, will they pay for the refrigerator? Write a multiplication and a division sentence to support your answer.

_____

_____

$984

17. Test Practice Heather bought a package of construction paper that holds 961 pieces with 31 different colors. If there are the same number of pieces per color, how many pieces of each color are there?

Ⓐ 992   Ⓒ 31

Ⓑ 30   Ⓓ 40

Lesson 4
Adjust Quotients

ESSENTIAL QUESTION
What strategies can I use to divide by a two-digit divisor?

When you estimate which digit to place in the quotient, your estimate might be too small or too large. So, you need to adjust the quotient.

Math in My World

Watch | Tutor

Example 1

During lunch, there were 144 students in the cafeteria. The cafeteria has a total of 16 tables. How many students can sit at each table?

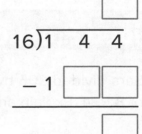

Let *s* represent the number of students at each table. Write an equation to find the value of *s*.

_____ ÷ _____ = *s*

1 Estimate by using compatible numbers. 140 ÷ 20 = _____

2 Try the estimate.

```
     □
16)1 4 4
  − 1 1 □
  _____
    □ □
```

Since 32 > 16, the estimated digit is too low.

3 Adjust. Try 8.

```
     □
16)1 4 4
  − 1 □ □
  _____
    □ □
```

Since 16 = 16, the estimated digit is too low.

4 Adjust again. Try 9.

```
     □
16)1 4 4
  − 1 □ □
  _____
    □
```

So, 144 ÷ 16 = _____. Since *s* = _____, _____ students can sit at each table.

Check for Reasonableness _____ ≈ _____

Example 2

Find 1,252 ÷ 32.

1 Estimate by using compatible numbers.

$1,252 \div 32$

$1,200 \div 30 =$ _____

2 Try the estimate.

$$32\overline{)1,\ 2\ 5\ 2}$$

Since 128 > 125, the estimated digit is too high.

3 Adjust. Try 3.

$$32\overline{)1,\ 2\ 5\ 2}$$

29 < 32
Continue dividing.

4 Bring down the 2 ones. Try 9.

$$32\overline{)1,\ 2\ 5\ 2}$$

R

4 < 32

So, $1,252 \div 32 =$ _____ R _____.

Check for Reasonableness _____ R _____ ≈ _____

Guided Practice

1. Sara divided 306 by 34 and got a quotient of 8 R34. Explain and correct her error.

_____

_____

_____

_____

Explain how you know when a digit you try in the quotient is too small.

Independent Practice

Divide. Check each answer.

2. $1{,}272 \div 53 =$ _____

3. $548 \div 62 =$ _____

4. $5{,}243 \div 71 =$ _____

5. $115 \div 23 =$ _____

6. $1{,}728 \div 72 =$ _____

7. $183 \div 19 =$ _____

8. $57\overline{)413}$

9. $34\overline{)242}$

10. $64\overline{)2{,}712}$

Processes &Practices **2** **Use Algebra** **Divide to find the variable in each equation.**

11. $328 \div 41 = m$

12. $4{,}536 \div 81 = w$

13. $735 \div 15 = x$

$m =$ _____

$w =$ _____

$x =$ _____

Problem Solving

14. Sheila arranged a total of 680 chairs for a school assembly. If she placed an equal amount of chairs in 20 rows, how many chairs are in each row?

_____

15. Given the area of a rectangle is 208 square inches, and the length is 26 inches, find the width.

_____

16. A crew went net fishing to catch shrimp. They caught 486 shrimp in 54 minutes. How many shrimp did they catch per minute? Find the unknown number in the equation $486 \div 54 = s$.

_____

Brain Builders

17. **Processes &Practices** **3** **Find the Error** Emma estimated the first digit in the quotient of $2,183 \div 42$ as 4. She adjusted the quotient to 3. What did she do wrong? Explain how Emma can avoid her mistake.

$$\begin{array}{r} 4 \\ 42\overline{)2,183} \\ -168 \\ \hline 50 \quad 50 > 42 \\ \text{I'll try 3.} \end{array}$$

_____

_____

18. **Building on the Essential Question** How can I adjust a quotient to solve a division problem? Give an example.

_____

_____

_____

Name

MY Homework

Homework Helper

Need help? connectED.mcgraw-hill.com

Find 238 ÷ 62.

1 Estimate by using compatible numbers.

$$238 \div 62$$

$$240 \div 60 = 4$$

2 Try the estimate.

$$
\begin{array}{r}
4 \\
62\overline{)238} \\
-248 \\
\end{array}
$$

Since 248 > 238,
the estimated digit
is too high.

3 Adjust. Try 3.

$$
\begin{array}{r}
3 \text{ R52} \\
62\overline{)238} \\
-186 \\
\hline
52 \\
\end{array}
$$

52 < 62

So, 238 ÷ 62 = 3 R52.

Check for Reasonableness 3 R52 ≈ 4

Practice

Divide. Check each answer.

1. 48)1,261

2. 86)1,204

3. 428 ÷ 61 = _____

Algebra Divide to find the variable in each equation.

4. $140 \div 28 = t$

$t = \underline{\hspace{1.5cm}}$

5. $2,075 \div 83 = c$

$c = \underline{\hspace{1.5cm}}$

6. $531 \div 59 = n$

$n = \underline{\hspace{1.5cm}}$

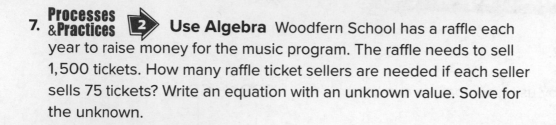

Brain Builders

7. **Processes &Practices** **2** **Use Algebra** Woodfern School has a raffle each year to raise money for the music program. The raffle needs to sell 1,500 tickets. How many raffle ticket sellers are needed if each seller sells 75 tickets? Write an equation with an unknown value. Solve for the unknown.

_____

8. An electronics store prices 3 dozen video games at a total of $1,476. If each video game costs the same amount, what is the total cost for 4 dozen video games? Show your work by writing equations with variables for the unknown values.

_____

_____

9. The Rodriquez family is taking a 1,311-mile train ride. If the train travels 57 miles per hour, how many hours will the ride last? Model with two equations using variables for the unknown values. Write one using compatible numbers and one using actual numbers.

_____

_____

10. **Test Practice** Alaska has the longest coastline in the United States. Traveling at 62 miles per hour, how many hours would it take to travel along the Pacific Coast?

Ⓐ 17 hours

Ⓒ 90 hours

Ⓑ 89 hours

Ⓓ 100 hours

Alaska Coastline	
Coast	**Miles**
Pacific	5,580
Arctic	1,060

Lesson 5
Divide Greater Numbers

ESSENTIAL QUESTION
What strategies can I use to divide by a two-digit divisor?

 Math in My World Watch Tutor

Example 1

A large city has a total of 22,500 students that ride a bus to school. There are 75 different schools within the city. How many students are dropped off at each school if an equal number of students are dropped off at each school?

Let s represent the number of students dropped off at each school. Write an equation to find the value of s.

_____ ÷ _____ = s

 Place the first digit.

$225 ÷ 75 = 3$

Write 3 in the quotient over the hundreds place.

 Multiply. $75 × 3 = 225$

Subtract. $225 − 225 = 0$

Compare. $0 < 75$

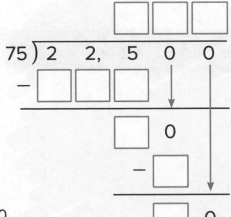

3 Divide the tens.

$0 ÷ 75 = 0$

$75 × 0 = 0$

$0 − 0 = 0$

$0 < 75$

4 Divide the ones.

$0 ÷ 75 = 0$

$75 × 0 = 0$

$0 − 0 = 0$

$0 < 75$

So, $22,500 ÷ 75 =$ _____ . Since $s =$ _____ , _____ students are dropped off at each school.

Example 2

Estimate the quotient of 46,534 and 152. Then divide. Is 36 a reasonable quotient? Explain.

Estimate 45,000 ÷ 150 = _____

 Place the first digit.

465 ÷ 152 ≈ 3

Write 3 in the quotient over the hundreds place.

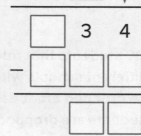

☐ ☐ ☐ R ☐ ☐

152) 4 6, 5 3 4

3 Divide the tens.

Ninety-three is not divisible by 152, so put a 0 in the quotient over the tens place.

2 Multiply.

152 × 3 = 456

Subtract. 465 − 456 = 9

Compare. 9 < 152

4 Divide the ones.

934 ÷ 152 ≈ 6

152 × 6 = 912

934 − 912 = 22

22 < 152

Check Since the estimate is _____ and the actual quotient is

_____ , a quotient of 36 is not reasonable.

Guided Practice

1. Find the missing number in the division problem below.

```
      1,■12
25 ) 47,800
   − 25 ↓
     22 8
    − 22 5↓
        30
      − 25↓
         50
       − 50
          0
```

Talk MATH

Explain how estimation can be used before, during, and after a division problem.

Independent Practice

Estimate. Then divide. Check for reasonableness.

2. 51)91,988

3. 17)14,637

4. 64)15,489

5. 36,712 ÷ 52 = _____

6. 43,803 ÷ 93 = _____

7. 26,208 ÷ 28 = _____

8. 42)25,435

9. 89)85,978

10. 783)52,056

Algebra Divide to find the variable in each equation.

11. 39,788 ÷ 812 = y

12. 25,696 ÷ 352 = g

13. 36,557 ÷ 263 = d

y = _____

g = _____

d = _____

Problem Solving

14. A farmer plows a corn field in the shape of a rectangle that has an area of 15,840 square yards. If the length of the field is 132 yards, what is the width of the field?

_____

15. An average person speaks 35,000 words in one week. Does the average person speak more or less than 2,500 words per day? Find the unknown number in the equation $35{,}000 \div 7 = w$.

_____

Brain Builders

16. **Processes &Practices** **2** **Use Number Sense** The athletic department raised $14,533 to buy new football uniforms. If each uniform costs $258, how many uniforms can they buy? Explain how you interpreted the remainder. How much more money do they need to buy one more uniform?

_____

17. **Processes &Practices** **3** **Draw a Conclusion** Find the unknown number in the equation $30{,}672 \div q = 852$. Explain how you found the unknown.

_____

18. **Building on the Essential Question** How can I divide greater numbers using a standard procedure?

_____

_____

Name _____

MY Homework

Lesson 5

Divide Greater Numbers

Homework Helper

Need help? ⟋ **connectED.mcgraw-hill.com**

At a recent rock concert, **$28,440 was made from front-row ticket sales. If the cost of a single ticket was $72, how many people purchased front-row tickets?**

Find 28,440 ÷ 72.

Estimate 28,000 ÷ 70 = 400

1 Place the first digit.

284 ÷ 72 ≈ 3

Write 3 in the quotient over the hundreds place.

```
          395
 72)28,440
    −216↓
      6 84
    − 6 48↓
         360
       − 360
           0
```

2 Multiply. 72 × 3 = 216

Subtract. 284 − 216 = 68

Compare. 68 < 72

3 Divide the tens.

684 ÷ 72 ≈ 9

72 × 9 = 648

684 − 648 = 36

36 < 72

4 Divide the ones.

360 ÷ 72 = 5

72 × 5 = 360

360 − 360 = 0

0 < 72

So, 395 people purchased front-row tickets.

Check Since the estimate is 400, the answer of 395 is reasonable.

Practice

Estimate. Then divide. Check for reasonableness.

1. 21,312 ÷ 36 = _____

2. 76,912 ÷ 92 = _____

3. 26,878 ÷ 89 = _____

Copyright © McGraw-Hill Education

Problem Solving

4. Given the area of a rectangle is 14,628 square millimeters, and the width is 12 millimeters, find the length.

_____

5. Turtles and tortoises have long life spans. A tortoise can live for 54,750 days. How many years can a tortoise live? (_hint_: 365 days = 1 year)

_____

Brain Builders

6. **Processes & Practices** **2** **Use Algebra** The new baseball stadium holds 64,506 people. There are 26 gates where people enter the ballpark. The same number of people entered each gate. How many people entered the first gate? Write an equation using a variable for the unknown and solve.

_____

_____

7. **Test Practice** The 38,483 employees at a manufacturing company must attend a meeting about finances. The meeting room holds 315 people. How many meetings does the manufacturing company need to schedule?

Ⓐ 122   Ⓒ 124

Ⓑ 123   Ⓓ 125

Lesson 6
Problem-Solving Investigation
STRATEGY: Solve a Simpler Problem

Learn the Strategy

Miguel earns $50 each week. He spends $10 each week and saves the remaining amount. How many weeks will it take until he has saved more than $300?

1 Understand

What facts do you know?

I know that Miguel earns _____ each week, but spends _____ of it.

What do you need to find?

I need to find how long it will take Miguel to save _____ .

2 Plan

I can solve the problem by solving a simpler problem.

3 Solve

Miguel saves $50 − $10, or _____ each week.

Divide 300 ÷ 40. Since there is a remainder,

Miguel will need to save for at least _____ weeks.

$$\begin{array}{r} 7\ \text{R20} \\ 40\overline{)300} \\ -\ 280 \\ \hline 20 \end{array}$$

4 Check

Is my answer reasonable? Explain.

Yes. _____ × 8 = _____ , and _____ is greater than $300.

Copyright © McGraw-Hill Education

Melissa estimates that she watched 104 movies over the past year. On average, about how many movies does she watch per month? (*hint:* 1 year = 52 weeks)

 Understand

What facts do you know?

_____

What do you need to find?

_____

2 Plan

_____

3 Solve

4 Check

Is my answer reasonable? Explain.

_____

Apply the Strategy

Solve each problem by solving a simpler problem.

1. Mr. Santiago has a flight from New York to Paris that covers a distance of 3,640 miles in 7 hours. If the plane travels at the same speed per hour, how many miles will it have traveled after 4 hours?

_____

2. Josh watches 720 television shows in one year. If he watches the same number of shows each month, how many shows does he watch in 5 months?

_____

3. The football team is raising money to have a new turf field installed. The cost of the turf field is $48,780. The team has 18 months to raise the money. If they raise an equal amount each month, how much will they have raised after one year?

_____

Brain Builders

4. **Processes &Practices** 🔼 **Keep Trying** The Clearview Window Washing Company has a contract to wash 3,082 windows on a 23-story building. The company will wash the windows on the first three floors on the first day. If there are the same number of windows on each floor, how many windows will the company wash on the first day? Explain the steps to solve.

_____

5. Mr. Thomas is delivering bricks to a construction site. He delivers the same number of bricks with each delivery. The builder has ordered 3,096 bricks and it will take 8 trips to deliver all the bricks. How many bricks will Mr. Thomas have delivered after 5 trips? How many bricks are left to be delivered?

_____

Use any strategy to solve each problem.

- Solve a simpler problem.
- Determine extra or missing information.
- Make a table.
- Use the four-step plan.

6. An aquarium at a pet store has 18 Black Neon Tetra fish in it. A customer buys 12 Black Neon Tetra fish at the same time the store clerk adds 7 more Black Neon Tetra fish to the tank. How many Black Neon Tetra fish are in the aquarium now?

7. Kolby read 108 books in one year. If he reads the same number of books each month, how many books does he read in 5 months?

8. Mr. Reyes baked 4 batches of muffins for his class. Each batch had 12 muffins. If Mr. Reyes has 24 students, how many muffins will each student receive?

9. The quotient of two numbers is 20. Their sum is 84. What are the 2 numbers?

10. **Processes &Practices** 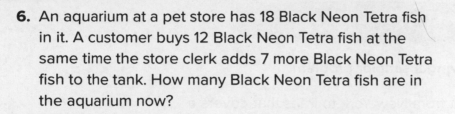 **Model Math** Charity and her friend each want to buy a piece of pizza, a drink, and an ice cream cone. Charity has $10 to pay for her and her friend's meal. Does she have enough money? Explain.

Menu	
Pizza	$3.00
Drink	$1.00
Ice cream cone	$2.00

MY Homework

Homework Helper

Need help? connectED.mcgraw-hill.com

Leslie wants to plant flowers in a rectangular garden that has a length of 6 feet and a width of 5 feet. If a tree takes up a square that is 2 feet on each side in the middle of the garden, how much area is left to plant the flowers?

1 Understand

What facts do you know?

I know the length and width of the garden and the space the tree takes up.

What do you need to find?

I need to find the area of the garden remaining for flowers.

2 Plan

You can solve the problem by solving a simpler problem.

3 Solve

Find the area of the entire garden. (*hint*: Area = length × width)

$6 \times 5 = 30$ square feet

Find the area of the square the tree takes up.

$2 \times 2 = 4$ square feet

Subtract the area of the tree from the area of the garden.

$30 - 4 = 26$ square feet

So, Leslie can plant 26 square feet of flowers.

4 Check

Is my answer reasonable? Explain.

area of flower garden + area of tree = total area

26 square feet + 4 square feet = 30 square feet

Problem Solving

Solve each problem by solving a simpler problem.

1. Harold stores 10,656 pounds of grain for his 24 cattle. Each cow eats about 12 pounds of grain each day. How many days would the food supply last?

2. **Processes &Practices** **1** **Make Sense of Problems** The Pave the Way Company is installing a new walkway. The installer uses 13 paving stones for every foot of walkway. He needs 117 paving stones to complete the project and has used 78 paving stones so far. How many more feet of paving stones will he need to install to finish?

Brain Builders

3. Harvey collects rare marbles. He has 125 marbles in Collection 1, 36 fewer marbles in Collection 2, and 53 more marbles than Collection 1 in Collection 3. Each display tray holds an equal number of marbles. If he uses 14 trays to hold all of the marbles, how many marbles will he display on 3 trays? What simpler problems did you chose to solve first? Explain.

4. Noelle's rectangular deck has a length of 18 feet and a width of 16 feet. If a hot tub takes up a square that is 4 feet on each side in the middle of the deck, how much area is left of the deck? Explain how you know.

5. Hilary wants to go on the Latin Club trip to Italy. It will cost $2,730 for the trip. The trip is 30 weeks away and she wants to make equal weekly payments. How much money altogether does Hilary need to pay at the end of week 8? Explain how to find how much money Hilary will still owe.

Fluency Practice

Divide.

1. 12)‾4‾3‾2‾   **2.** 40)‾8‾5‾0‾   **3.** 29)‾4‾9‾3‾   **4.** 18)‾5‾3‾5‾

5. 32)‾1‾0‾5‾   **6.** 62)‾8‾9‾7‾   **7.** 72)‾5‾0‾4‾   **8.** 53)‾6‾8‾9‾

9. 21)‾1‾2‾6‾   **10.** 37)‾4‾5‾2‾   **11.** 76)‾6‾8‾4‾   **12.** 51)‾2‾0‾9‾

13. 67)‾9‾3‾8‾   **14.** 41)‾8‾6‾8‾   **15.** 34)‾6‾4‾6‾   **16.** 15)‾8‾7‾9‾

17. 10)‾3‾2‾9‾   **18.** 98)‾7‾8‾4‾   **19.** 39)‾9‾7‾5‾   **20.** 47)‾6‾6‾0‾

Fluency Practice

Divide.

1. $13\overline{)32{,}708}$
2. $76\overline{)81{,}928}$
3. $54\overline{)65{,}775}$
4. $83\overline{)72{,}628}$

5. $23\overline{)22{,}890}$
6. $61\overline{)61{,}427}$
7. $90\overline{)65{,}881}$
8. $38\overline{)89{,}148}$

9. $15\overline{)92{,}460}$
10. $43\overline{)27{,}998}$
11. $27\overline{)61{,}752}$
12. $87\overline{)50{,}286}$

13. $216\overline{)16{,}200}$
14. $557\overline{)60{,}156}$
15. $318\overline{)80{,}779}$
16. $861\overline{)89{,}550}$

17. $396\overline{)88{,}704}$
18. $221\overline{)82{,}889}$
19. $562\overline{)55{,}103}$
20. $902\overline{)92{,}004}$

Vocabulary Check

Match each word to its definition. Write your answers on the lines provided.

1. **compatible numbers** _____

   A. To find the approximate value of a number.

2. **dividend** _____

   B. A number that is being divided.

3. **divisor** _____

   C. Numbers in a problem that are easy to compute mentally.

4. **quotient** _____

   D. The number that is left after one whole number is divided by another.

5. **remainder** _____

   E. The number that divides the dividend.

6. **round** _____

   F. The result of a division problem.

Concept Check

Estimate. Show how you estimated.

7. $74\overline{)634}$

8. $49\overline{)311}$

9. $38\overline{)409}$

Divide. Check each answer.

10. $32\overline{)928}$

11. $23\overline{)345}$

12. $53\overline{)753}$

13. $926 \div 71 = $ _____

14. $126 \div 17 = $ _____

15. $478 \div 93 = $ _____

Estimate. Then divide. Check for reasonableness.

16. $85,120 \div 76 = $ _____

17. $54,184 \div 26 = $ _____

18. $25,600 \div 25 = $ _____

19. $61\overline{)37,520}$

20. $41\overline{)16,859}$

21. $53\overline{)75,578}$

Problem Solving

22. Tito purchased 11 tickets to a baseball game for $374. About how much did one ticket cost?

_____

23. Tonya saved $564 in one year. If she saved the same amount each month, how much did Tonya save in two months?

_____

Brain Builders

24. Bennett has 107 photographs to place in the school yearbook. He will put the same number of photos on each of the 13 pages. If he can put 4 pictures in a row, how many photographs will not be placed in the yearbook? Explain.

_____

_____

25. Trevon is making payments on a car that costs $26,555. He makes 36 equal payments. If he rounds the equal payments up to the nearest whole dollar, about how much will he overpay after 36 months? Show your work to justify your answer.

_____

_____

26. Test Practice A large farm tractor costs $98,424. If a farmer makes 36 equal payments, how much is each payment?

Ⓐ $2,734   Ⓒ $2,624

Ⓑ $2,714   Ⓓ $2,604

Reflect

Use what you learned about dividing whole numbers to complete the graphic organizer.

Write the Example

Real-World Example

ESSENTIAL QUESTION

What strategies can I use to solve division problems with whole numbers?

Estimate

Vocabulary

Now reflect on the ESSENTIAL QUESTION Write your answer below.

_____

_____

Performance Task

Constructing Frames for Habitat Models

Mr. Cruz is making rectangular frames for his students' Earth Science project. Each frame encloses an area of 3,456 square inches. An example of the frame is shown below:

48 inches

Show all your work to receive full credit.

Part A

What is the length of the frame? Explain.

Part B

Mr. Cruz goes to a supply store and finds that boards are only sold in lengths of 5 feet and 7 feet. Are these boards long enough for the project? Explain.

Part C

Mr. Cruz has a grant for $1,350. He needs to make 18 habitat models. How much can he spend on supplies for each habitat model?

Part D

Fill in the chart for how many 5-foot boards and 7-foot boards Mr. Cruz will need for the habitat models and fill in the amount of wood left over. Explain.

Board Length	Boards Needed for 18 Frames	Total Inches Left Over

ESSENTIAL QUESTION

How can I use place value and properties to add and subtract decimals?

Let's Explore Technology!

Watch

Watch a video!

MY Chapter Project

Food Drive

Get started!

(Group 1) Create a poster listing food items needed for a local food pantry.

(Group 2) Use the Internet, grocery ads, and other resources to estimate the dollar value of each item.

(Group 3) Determine the number of each item your class will collect.

Record it!

During the food drive, record the donations on the poster.

Wrap it up!

1. Assess whether your class reached its goal and estimate the total dollar value of the items collected.

2. Below, write a summary of the project.

3. On a separate sheet of paper, create a bar graph or pictograph to display the results.

_____

_____

_____

_____

_____

_____

_____

_____

Am I Ready?

Name the place-value position of each highlighted digit.

1. 52 _____

2. 138 _____

3. 4.3 _____

4. 901 _____

5. 1.216 _____

6. 2,785 _____

Round each number to the highlighted place.

7. 19 = _____

8. 681 = _____

9. 735 = _____

10. 3,705 = _____

11. 106,950 = _____

12. 5,750 = _____

Add.

13. 38 + 16 = _____

14. 151 + 218 = _____

15. 260 + 398 = _____

16. 235 + 68 = _____

17. The Pham family and the Weber family have many pets. How many more pets does the Pham family have than the Weber family?

_____

Pets	
Pham	**Weber**
3 dogs	2 dogs
1 cat	3 gerbils
6 fish	1 turtle

How Did I Do?

Shade the boxes to show the problems you answered correctly.

1	2	3	4	5	6	7	8	9	10	11	12	13	14	15	16	17

Online Content at connectED.mcgraw-hill.com

MY Math Words

Vocab
abc

Review Vocabulary

greater than (>)   less than (<)   equal to (=)

Making Connections

Compare the numbers in each row. Use the review vocabulary to compare the two numbers in each row using >, <, or =.

Greater Than, Less Than, or Equal To

57.2		57.02
12.01		12.1
12.6		12.60
24.56		24.5
6.99		6.89

Describe how you used place value to complete the chart.

....

....

....

MY Vocabulary Cards

Processes & Practices

Lesson 5-7

Associative Property of Addition

$(15.19 + 25.05) + 88 =$
$15.19 + (25.05 + 88)$

Lesson 5-7

Commutative Property of Addition

$15.9 + 8.42 = 8.42 + 15.9$

Lesson 5-7

Identity Property of Addition

$12.7 + 0 = 12.7$

Lesson 5-10

inverse operations

$1.73 - 0.87 = 0.86$

$0.87 + 0.86 = 1.73$

Ideas for Use

- Write a tally mark on each card every time you read the word in this chapter or use it in your writing. Challenge yourself to use at least 10 tally marks for each card.

- Use the blank cards to write your own vocabulary cards.

The order in which numbers are added does not change the sum.

Commutative means "something that involves substitution." How does this definition relate to the Commutative Property?

_____

_____

The way in which numbers are grouped does not change the sum.

Compare the example on this card with the example on the Commutative Property card. What differences do you notice?

_____

_____

Operations that undo each other.

What are the inverse operations for addition and division?

_____

The sum of any number and 0 equals the number.

Explain how one of the three properties in this chapter can help you add mentally.

_____

_____

MY Foldable

FOLDABLES® Follow the steps on the back to make your Foldable.

Estimate

1.29 + 1.55 = _____

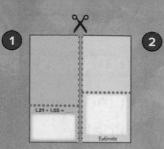

① ②

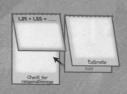

③

Check for Reasonableness

Add

Name ....

Lesson 1
Round Decimals

ESSENTIAL QUESTION
How can I use place value and properties to add and subtract decimals?

Recall that numbers with digits in the tenths place, hundredths place, or beyond are called decimals. When you round a decimal, you find its approximate value.

 Math in My World

Example 1

Andrew's laptop has a processor with a speed of 2.8 gigahertz. Round the processing speed of the laptop to the nearest whole number.

Use a number line to round 2.8 to the nearest whole number.

 Draw 10 equal increments between 2 and 3 on the number line.

2   3

 Place a dot at 2.8 and label it.

Determine if 2.8 is closer to 2 or 3. ◄——

There are 2 equal increments between 2.8 and 3. There are 8 equal increments between 2 and 2.8.

2.8 is closer to _____.

So, round 2.8 to _____.

Key Concept Rounding Decimals

- Underline the digit in the place to which you want to round.
- Look at the digit to its right. If the digit is 4 or less, keep the underlined digit. If the digit is 5 or greater, round the underlined digit up to the next greater digit.
- Drop the digit to the right of the underlined digit.

Example 2

Round 46.73 to the nearest tenth.

1 Underline the digit in the tenths place, _____ .

46.73

2 Look at the digit to the right of 7.

_____ 3 is _____ than 5

3 So, keep the underlined digit. Drop the digit to its right.

46.73 ⟶ 46._____

So, 46.73 rounds to _____ .

Guided Practice

1. Round 8.74 to the nearest one. Underline the digit in the ones place.

   8.74

   Look at the _____, the digit to the right of 8.

   So, 8.74 rounds to _____ .

Explain how to round 74.685 to the nearest hundredth.

Independent Practice

Round each decimal to the place indicated.

2. 5.476; hundredths

3. 983.625; hundredths

4. 28.6; ones

_____

_____

_____

5. 4.35; tenths

6. 110.079; hundredths

7. 67.142; ones

_____

_____

_____

8. 1.8; ones

9. 7.358; hundredths

10. 48.32; ones

_____

_____

_____

11. 9.045; tenths

12. 19.25; ones

13. 8.17; tenths

_____

_____

_____

Problem Solving

14. What is the length of the 10-dollar bill to the nearest whole number?

|←——— **15.6 cm** ———→|

15. The weight of a new touch screen device is 1.6 pounds. What is the weight to the nearest whole number?

Brain Builders

16. Processes &Practices **1** **Keep Trying** Write two different numbers that when rounded to the nearest tenth will give you 18.3. Write the first number to the hundredths and the second number to the thousandths.

17. Processes &Practices **2** **Use Number Sense** Explain what happens when you round 9,999.999 to any place.

Write a number that rounds to 5,000 when rounded to any place.

18. ❓ Building on the Essential Question In what real-world situations would you want to round a number? List 3 examples.

MY Homework

Homework Helper

Need help? ↗ connectED.mcgraw-hill.com

An ice sheet that covers most of Antarctica is about 1.34 miles thick. To the nearest tenth of a mile, how thick is the ice?

Round 1.34 to the nearest tenth.

1 The digit in the tenths place is 3.

2 Look at the digit to the right of 3.
4 is less than 5

1.34

3 So, keep the 3 and drop the digit to its right, 4.

1.34 ⟶ 1.3

So, the thickness of the ice rounded to the nearest tenth is 1.3 miles.

Practice

Round each decimal to the place indicated.

1. 5.476; hundredths

2. 4.35; tenths

3. 1.8; ones

4. 0.79; ones

5. 1.049; hundredths

6. 17.92; tenths

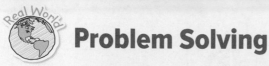

Problem Solving

Use the information in the table to solve Exercises 7–9. Round each number to the place indicated.

Place	Area (square miles)
Florida	65,754.59
Georgia	59,424.77
Alabama	52,419.02
South Carolina	32,020.20

7. What is the area of Florida rounded to the nearest tenth?

_____

8. What is the area of South Carolina rounded to the nearest tenth?

_____

9. What is the area of Georgia rounded to the nearest square mile?

_____

Brain Builders

10. The African bush elephant weighs between 4.4 tons and 7.7 tons. What are the least and greatest weights of 3 elephants, rounded to the nearest ton?

_____

11. **Processes &Practices** **6** **Be Precise** The price of a gallon of milk, rounded to the nearest dollar, is $4.00. Between what two amounts could the actual price be?

_____

12. **Test Practice** Holly bought blank CDs and spent $34.57. What is that amount rounded to the nearest dollar?

Ⓐ $35   Ⓒ $34.50

Ⓑ $34.60   Ⓓ $34

Lesson 2
Estimate Sums and Differences

ESSENTIAL QUESTION ?
How can I use place value and properties to add and subtract decimals?

One way to estimate is to use rounding. If you round numbers to a lesser place value, you are likely to get an estimate that is closer to the exact answer.

 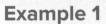 Math in My World Watch ▶ Tutor 💬

Example 1

Hadassah used a digital thermometer to find the morning temperature and afternoon temperature. She found the morning temperature was 31.3°F and the afternoon temperature was 37.6°F. Estimate the difference in average temperatures.

One Way Round to the nearest ten.	**Another Way** Round to the nearest one.
Round 37.6 to the nearest ten.	Round 37.6 to the nearest one.
37.6 ⟶ _____	37.6 ⟶ _____
Round 31.3 to the nearest ten.	Round 31.3 to the nearest one.
31.3 ⟶ _____	31.3 ⟶ _____
Subtract.	Subtract.
_____ − _____ = _____	_____ − _____ = _____

The difference is about _____ °F or about _____ °F.

The actual difference is 6.3°F. So, rounding to the nearest _____ gave the more accurate estimate.

Example 2

Estimate 5.26 + 1.93 by rounding to the nearest one.

5.26 → _____

1.93 → _____

Add.

_____ + _____ = _____

The sum is about _____.

Guided Practice

Round each decimal to the nearest one. Then add or subtract.

1. 2.8 + 1.3

   Round to the nearest one.

   2.8 → _____

   1.3 → _____

   Add.

   _____ + _____ = _____

   So, 2.8 + 1.3 is about _____.

Talk MATH

Describe a real-world example when it might be appropriate to estimate rather than get the exact answer.

2. 5.98 − 1.03

   Round to the nearest one.

   5.98 → _____

   1.03 → _____

   Subtract.

   _____ − _____ = _____

   So, 5.98 − 1.03 is about _____.

Independent Practice

Round each decimal to the nearest one. Then add or subtract.

3. $10.08 + 5.6 =$ _____

4. $10.4 + 32.8 =$ _____

5. $\$42.01 - \$5.92 =$ _____

6. $75.2 + 82.3 =$ _____

7.
$$\begin{array}{r} 1.509 \\ + 3.106 \\ \hline \end{array}$$

8.
$$\begin{array}{r} 8.058 \\ - 3.181 \\ \hline \end{array}$$

9.
$$\begin{array}{r} 3.872 \\ + 1.249 \\ \hline \end{array}$$

Round each decimal to the nearest ten. Then add or subtract.

10. $23.78 + 10.45 =$ _____

11. $83.69 - 55.41 =$ _____

12.
$$\begin{array}{r} 37.58 \\ - 21.25 \\ \hline \end{array}$$

13.
$$\begin{array}{r} 32.56 \\ + 6.7 \\ \hline \end{array}$$

14.
$$\begin{array}{r} 25.21 \\ - 12.47 \\ \hline \end{array}$$

Problem Solving

Solve Exercises 15–17 by rounding to the nearest one.

15. The weights of Marisa and Toni's televisions are shown in the table. About how much more does Marisa's television weigh than Toni's?

Student	Television Weight (lb)
Marisa	52.7
Toni	43.9

_____

16. **Processes &Practices** 4 **Model Math** Sophia has $20. She buys a hair band for $3.99, gum for $1.29, and a brush for $6.75. Not including tax, estimate how much change she should receive. Show your work.

_____

17. Malcolm buys a taco for $1.79 and milk for $1.29. About how much money does he spend? Show your work.

_____

Brain Builders

18. **Processes &Practices** 3 **Find the Error** Kim wants to estimate 5.494 + 1.108 by first rounding to the nearest hundredth. Explain her error and correct it. Then find the estimated sum.

$$\begin{array}{r} 5.494 \\ + 1.108 \end{array} \longrightarrow \begin{array}{r} 5.50 \\ + 1.11 \end{array}$$

_____

19. **Building on the Essential Question** When is estimating an effective tool? List two examples.

_____

_____

_____

Name ....

Homework Helper

Need help? ⟋ **connectED.mcgraw-hill.com**

The results of a recent skateboard competition are shown. About how many more points did Steamer have than Dal Santo?

Athlete	Points
Elissa Steamer	87.83
Marisa Dal Santo	81.50
Amy Caron	80.00

Round each decimal to the nearest ten.
Then subtract.

Steamer ⟶ 87.83 ⟶ 90
Dal Santo ⟶ 81.50 ⟶ − 80
                                        ‾‾‾‾‾‾
                                          10

So, Steamer scored about 10 more points than Dal Santo.

Practice

Round each decimal to the nearest one. Then add or subtract.

1. $3.85
       − $2.17

2.   52.85
       − 9.09

3.   19.83
       + 9.93

4. 3.872 + 2.409 = _____

5. 9.086 − 2.419 = _____

Problem Solving

Solve Exercises 6–8 by rounding to the nearest one.

6. The table shows the average speeds of two airplanes in miles per hour. About how much faster is the Foxbat than the Hawkeye? Show your work.

Plane	Speed (mph)
Hawkeye	375.52
Foxbat	1,864.29

7. Aluminum and tin are both metals. The standard atomic weight for aluminum is 26.98. The standard atomic weight for tin is 118.71. Estimate the difference between the standard atomic weights of these two metals. Show your work.

8. Lorena and her cousin are fishing at the lake. They caught two largemouth bass. One fish weighs 71.27 ounces and the other fish weighs 38.86 ounces. Estimate the total weight of the two fish. Show your work.

Brain Builders

9. **Processes &Practices** **4** **Model Math** The table shows the lengths of four trails at a horseback riding camp. The estimated length of all four trails is 12 miles. Find a possible length for trail D. Show your work.

Trail	A	B	C	D
Length (mi)	2.6	1.8	4.2	

10. **Test Practice** Mr. Dixon bought a whiteboard that was on sale for $1,989.99. The regular price was $2,499.89. Select the true statement.

Ⓐ The sale price was about half the regular price.

Ⓑ The regular price was about $1,000 more than the sale price.

Ⓒ Mr. Dixon saved about $500.

Ⓓ Mr. Dixon saved about $1,000.

Lesson 3
Problem-Solving Investigation
STRATEGY: Estimate or Exact Answer

Learn the Strategy Watch ▶ Tutor 💬

Mrs. Trainer needs to purchase a classroom response clicker and a new calculator. A clicker costs $31.99 and a calculator costs $11.75. About how much money will it cost to buy both?

1 Understand

What facts do you know?

Mrs. Trainer needs to buy a clicker for $31.99 and a calculator for $11.75.

What do you need to find?

about how much it will _____ to buy the clicker and calculator

2 Plan

You can _____ the cost of the clicker and calculator.

3 Solve

Round the costs of the clicker and calculator to the nearest whole dollar.

$31.99 ⟶ _____ $11.75 ⟶ _____

Next, add to find the estimated total cost.

_____ + _____ = _____

So, Mrs. Trainer will need to spend about _____ for the two items.

4 Check

Is my answer reasonable? Explain.

Estimate another way. $30 + $10 = _____

_____ ≈ _____

Madison's family drove to her grandparents' house. They drove 58.6 miles in the first hour, 67.2 miles in the second hour, and 60.5 miles in the third hour. They followed the same route to return home. About how far did Madison's family travel?

 Understand

What facts do you know?

_____

What do you need to find?

_____

 Plan

_____

_____

 **Solve**

 Check

Is my answer reasonable? Explain.

_____

_____

Apply the Strategy

Determine whether you need an estimate or an exact answer to solve each problem. Then solve.

1. A restaurant can make 95 dinners each night. The restaurant has been sold out for 7 nights in a row. How many dinners were sold during this week?

2. **Processes &Practices** ➊ **Plan Your Solution** A gardener has 35 feet of fencing to enclose the garden shown. About how much fencing will be left over after the garden is enclosed?

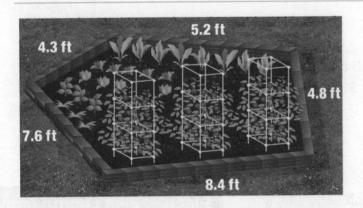

5.2 ft
4.3 ft
4.8 ft
7.6 ft
8.4 ft

Brain Builders

3. A family is renting a cabin for $59.95 a day for 3 days. They rent a canoe each day for $8.50. About how much will they pay?

4. Four friends ordered two pizzas. The cost of each pizza is $13.80. About how much will each friend pay if they share the cost equally?

5. Write a real world situation that would require an exact answer.

Review the Strategies

Use any strategy to solve each problem.

- Determine an estimate or exact answer.
- Make a table.
- Solve a simpler problem.
- Determine extra or missing information.

6. **Processes &Practices** **1** **Make Sense of Problems** Students at a high school filled out a survey. The results showed that out of 640 students, 331 speak more than one language. How many students speak only one language?

7. Pablo is dividing a dish of brownies for a party. The brownie mix costs $2.89. He cuts the brownies into squares that measure 3 inches by 3 inches. If the pan is 18 inches long and 12 inches wide, how many brownies did he cut?

8. Evita has 9 quarters, 7 dimes, and 5 nickels. Does she have enough money to buy a box of crayons for $3.25?

9. The fisherman with the longest fish wins a fishing competition. About how much longer is the first-place fish than the third-place fish?

Place	Fish Length (cm)
1st	68.7
2nd	59.8
3rd	58.2

10. Tammi is planning to buy a video game system for $310. Each month she doubles the amount she saved the previous month. If she saves $10 the first month, in how many months will Tammi have enough money to buy the video game system?

Copyright © McGraw-Hill Education

MY Homework

Homework Helper

Need help? connectED.mcgraw-hill.com

Each school raffle ticket costs $8.50. The Jordan family purchased 3 tickets. About how much did it cost the Jordan family to buy the raffle tickets?

1 Understand

What facts do you know?

Each ticket costs $8.50. The Jordan family purchased 3 tickets.

What do you need to find?

about how much it will cost for all 3 tickets

2 Plan

You can estimate to find the total cost.

3 Solve

Round $8.50 to the nearest whole dollar.

$8.50 ⟶ $9

Next, add to find the estimated total cost.

$9 + $9 + $9 = $27

So, the Jordan family spent about $27 to purchase the raffle tickets.

4 Check

Is my answer reasonable? Explain.

Estimate another way.

$10 + $10 + $10 = $30

$30 ≈ $27

Problem Solving

Determine whether you need an estimate or an exact answer to solve each problem. Then solve.

1. A library wants to buy a new computer that costs $989.99. So far, the library has collected $311.25 in donations. About how much more money does the library need to buy the computer?

2. On Friday, a museum had 185 visitors. On Saturday, there were twice as many visitors as Friday. On Sunday, 50 fewer people visited than Saturday. How many people visited the museum during these three days?

3. Tomás orders a meal that costs $7.89. Lisa's meal costs $9.05. About how much is the combined cost of their meals?

Brain Builders

4. **Processes &Practices** **1** **Make Sense of Problems** The ski team has a race in 3 hours and it is 127 miles away. The team drives about 48 miles per hour. Will they arrive at the race on time? Explain how you know.

5. At the beginning of last year, there were 368 students at the elementary school. By the beginning of this year, 72 of those students had moved to other school districts and 64 new students arrived. About how many students started the school year this year?

Vocabulary Check

Draw a line to match each word to its correct place in the number below.

1. ones   **2.** hundreds   **3.** hundredths   **4.** tens   **5.** tenths

$$1 \quad 2 \quad 3 . 4 \quad 5$$

Concept Check

Round each decimal to the place indicated.

6. 8.067; hundredths

_____

7. 178.03; tens

_____

8. 7.48; ones

_____

9. 26.138; tenths

_____

Round each decimal to the nearest one. Then add or subtract.

10. 32.9 + 17.2 = _____

11. 7.91 + 21.9 = _____

12. 48.21
      − 12.64

13. 107.14
      − 76.87

14.   6.239
      − 1.750

15.   3.902
      + 2.017

Problem Solving

For Exercises 16–19, determine whether you need an estimate or an exact answer. Then solve.

16. Book World receives 12 boxes of books. Each box contains 16 copies of the new best-seller, *Norton's Last Laugh*. How many copies of *Norton's Last Laugh* does the store receive?

_____

17. Students at Tuscan School filled out a survey. The survey showed that of 374 students, 195 speak a second language. How many students speak only one language?

_____

Brain Builders

18. Pensacola, Florida, gets an average of 64.28 inches of precipitation per year. Key West, Florida, gets an average of 38.94 inches per year. About how many more inches of precipitation does Pensacola get than Key West? Show your work.

_____

_____

19. It costs Marc $29.75 a week to feed his dog. About how much does it cost him to feed his dog for 3 weeks if the cost increases by $4.40 a week?

_____

_____

20. Test Practice The average price of a barrel of crude oil in a recent year was $65.37. Which estimated price is *incorrect*?

Ⓐ $65.00   Ⓒ $66.00

Ⓑ $65.40   Ⓓ $70.00

Name ....

Lesson 4
Hands On
Add Decimals Using Base-Ten Blocks

Build It **Tools**

Ruben downloaded two ringtones to his cell phone. One cost him $1.30 and the other one was discounted to $0.50. How much did it cost for both ringtones?

Find $1.30 + $0.50. Use base-ten blocks.

1 Model 1.3 and 0.5.

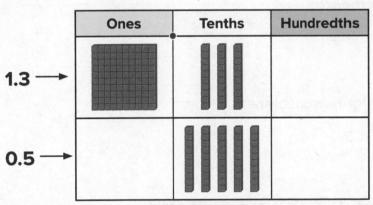

	Ones	Tenths	Hundredths
1.3 →			
0.5 →			

2 Combine the base-ten blocks. Draw the result.

How many tenths are there? _____

$1.30 + $0.50 = _____

So, Ruben spent _____ for both ringtones.

Helpful Hint
3 tenths + 5 tenths = 8 tenths

Check Use estimation to check for reasonableness.

$1 + $1 = $2 and _____ ≈ $2

Online Content at connectED.mcgraw-hill.com

Try It

Find 1.42 + 0.87. Use base-ten blocks.

 Model 1.42 and 0.87.

Ones	Tenths	Hundredths

Combine the base-ten blocks.
Since there are 12 tenths, you need to regroup. You can regroup 12 tenths as 1 whole and 2 tenths. Draw the result.

How many ones are there? _____

How many tenths are there? _____

How many hundredths are there? _____

So, 1.42 + 0.87 = _____.

Check Use estimation to check for reasonableness.

$$1 + 1 = 2 \text{ and } \underline{\hspace{1cm}} \approx 2$$

Talk About It

Processes & Practices ❷
1. **Reason** How is adding decimals similar to adding whole numbers?

_____

_____

_____

Practice It

Add. Use base-ten blocks. Draw each result in the table.

2. 0.3 + 0.4 = _____

Ones	Tenths	Hundredths

3. 2.4 + 0.5 = _____

Ones	Tenths	Hundredths

4. 1.52 + 0.37 = _____

Ones	Tenths	Hundredths

5. 3.71 + 1.53 = _____

Ones	Tenths	Hundredths

6. 1.2 + 0.9 = _____

Ones	Tenths	Hundredths

7. 2.71 + 0.45 = _____

Ones	Tenths	Hundredths

Apply It

Processes & Practices **5** **Use Math Tools** Nathan ran 2.35 miles on Tuesday and 3.15 miles on Thursday. What total distance did he run? Use base-ten blocks to help you solve.

_____

9. Valerie had $1.87 in her bank at home. She adds $2.67 in change. How much does she have now?

_____

10. **Processes & Practices** **3** **Find the Error** Tom used base-ten blocks to find 1.72 + 1.77. Find his mistake and circle the correct sum.

Ones	Tenths	Hundredths

Which is the correct sum?

2.29   3.14

3.29   3.49

_____

_____

Write About It

11. How can you use base-ten blocks to add decimals?

_____

_____

_____

Name

MY Homework

Homework Helper

Need help? connectED.mcgraw-hill.com

Jessica purchased a cup of hot chocolate for **$1.29** and a
granola bar for **$1.55.** How much did it cost for both items?

Find $1.29 + $1.55. Use base-ten blocks.

1 Model 1.29 and 1.55.

Ones	Tenths	Hundredths

2 Combine the base-ten blocks.

Since there are 14 hundredths, you need to regroup. You can regroup
14 hundredths as 1 tenth and 4 hundredths. The model shows the result.

Ones	Tenths	Hundredths

There are 2 ones, 8 tenths, and 4 hundreths.

So, the total for the hot chocolate and granola bar was $2.84.

Check Use estimation to check for reasonableness.

$1 + $2 = $3 and $2.84 ≈ $3

Practice

Add. Use base-ten blocks. Draw each result in the table.

1. 1.83 + 0.36 = _____

Ones	Tenths	Hundredths

2. 3.1 + 1.34 = _____

Ones	Tenths	Hundredths

Problem Solving

3. Lucas had 2.3 pounds of grapes left over from his class party. The class ate 1.9 pounds of the grapes. How many pounds of grapes did Lucas buy? Use models to find the sum.

_____

4. **Processes &Practices** **3** **Justify Conclusions** Janice wanted to download the following items for her cell phone: A ringtone for $3.10, a picture for $1.95, and a game for $2.05. She has $8. Not including tax, does Janice have enough money to buy all three items? Explain.

_____

5. Kelli had $0.55 in her pocket to buy a snack after school. Renee had $1.64 in her pocket. They decided to share a snack. How much can they spend on a snack altogether? Use models.

_____

Lesson 5

Hands On

Add Decimals Using Models

Copyright © McGraw-Hill Education   Ed-Imaging

ESSENTIAL QUESTION
How can I use place value and properties to add and subtract decimals?

Build It

Find 1.2 + 0.7. Use models to find the sum.

1 **Use 10-by-10 grids to model 1.2.**

To show 1.2, shade one whole 10-by-10 grid and two-tenths of a second grid.

Two-tenths is equal to _____ small squares.

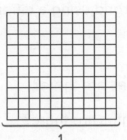

1

2 **Model 0.7.**

To show 0.7, shade seven-tenths of the second grid using a different color.

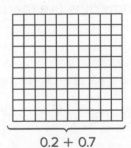

0.2 + 0.7

3 **Add.**

How many ones are there? _____

How many tenths are there? _____

So, 1.2 + 0.7 = _____.

Check Use estimation to check for reasonableness.

1 + 1 = 2 and _____ ≈ 2

Find 1.08 + 0.45. Use models.

 Model 1.08.

To show 1.08, shade one whole 10-by-10 grid and eight-hundredths of a second grid.

Eight-hundredths is equal to

_____ small squares.

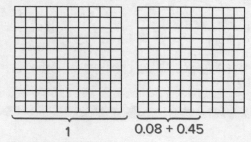

1   0.08 + 0.45

2 Model 0.45.

To show 0.45, shade forty-five hundredths of the second grid using a different color.

3 Add.

How many ones are there? _____

How many tenths are there? _____

How many hundredths are there? _____

So, 1.08 + 0.45 = _____.

> **Helpful Hint**
>
> 1 one, 5 tenths, and 3 hundredths is equal to 1.53.

Check Use estimation to check for reasonableness.

1 + 0.5 = 1.5 and _____ ≈ 1.5

Talk About It

1. Explain how to use grid paper to model 0.2 and 0.02. Describe any differences.

_____

_____

_____

Processes &Practices **4** **Model Math** What decimal is modeled by a
2. 10-by-10 grid with 82 small squares shaded? Explain.

_____

Practice It

Add. Shade the decimal models.

3. 2.46 + 1.13 = _____

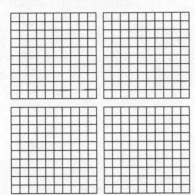

4. 2.05 + 1.87 = _____

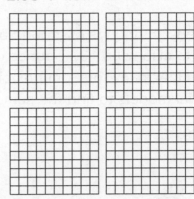

5. 2.91 + 1.8 = _____

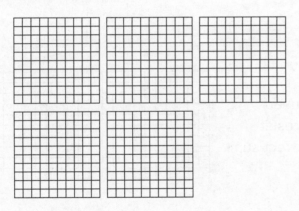

6. 1.34 + 1.15 = _____

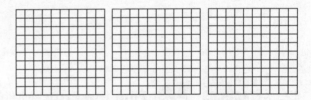

7. 1.74 + 0.36 = _____

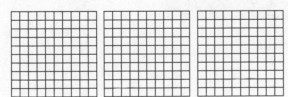

8. 2.05 + 1.12 = _____

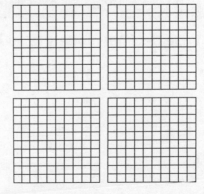

Apply It

9. The length of a nickel is 2.1 centimeters. What is the combined length of two nickels lying side by side? Use decimal models to find the sum.

_____

10. Shawn used 2.5 pounds of ground beef to make hamburgers for a cookout. He also used 1.32 pounds of ground beef to make spaghetti. How many total pounds of ground beef did Shawn use? Use decimal models to find the sum.

_____

11. **Processes &Practices** 5 **Use Math Tools** Marvin bought 0.6 pound of almonds and 1.73 pounds of cashews at the store. What is the total weight of the almonds and cashews he bought? Use decimal models to find the sum.

_____

12. **Processes &Practices** 3 **Find the Error** Melanie used decimal models to add 1.03 and 0.4. Her result was 1.07. Shade the models to find the correct sum.

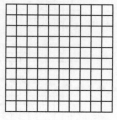

The correct sum is _____ .

Write About It

13. How can I use decimal models to add decimals?

_____

_____

_____

_____

MY Homework

Homework Helper

Need help? ↗ **connectED.mcgraw-hill.com**

Find 0.43 + 0.81. Use models.

1 **Use 10-by-10 grids to model 0.43.**

To show 0.43, shade forty-three hundredths of a 10-by-10 grid.

Forty-three hundredths is equal to 43 small squares.

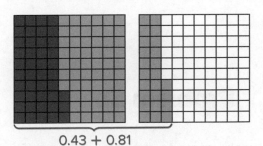

0.43 + 0.81

2 **Model 0.81.**

To show 0.81, shade an additional eighty-one hundredths using a different color.

3 **Add.**

There is 1 one.
There are 2 tenths.
There are 4 hundredths.

> **Helpful Hint**
>
> 1 one, 2 tenths, and 4 hundredths is equal to 1.24.

So, 0.43 + 0.81 = 1.24.

Check Use estimation to check for reasonableness.

$$0 + 1 = 1 \text{ and } 1.24 \approx 1$$

Practice

Add. Shade the decimal models.

1. 0.51 + 0.63 = _____

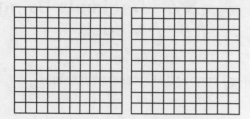

2. 1.93 + 1.73 = _____

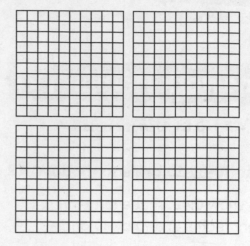

 Problem Solving

3. Norah wants to download two different applications on her cell phone. One used 2.42 megabytes and the other one used 1.76 megabytes. How many total megabytes of memory will Norah use to download the two applications? Draw decimal models to find the sum.

_____

4. **Processes &Practices** 4 **Model Math** Write a real-world problem that can be solved by adding 1.39 and 2.45 using decimal models. Then solve.

_____

_____

_____

5. Dorian feeds his dog 7.5 pounds of food in a week. He feeds his cat 3.75 pounds of food in a week. How many total pounds of food do his pets eat in a week? Draw decimal models to find the sum.

_____

Lesson 6
Add Decimals

ESSENTIAL QUESTION
How can I use place value and properties to add and subtract decimals?

To add decimals, line up the decimal points and add the digits in the same place-value position.

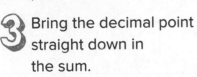 **Math in My World** Watch ▶ Tutor 💬

Example 1

Mr. Jacobson uses a digital scale to measure **44.2 milligrams** of sodium for a chemistry experiment. During the second experiment, he used **33.1 milligrams** of sodium. What is the total amount of sodium used?

Find 44.2 + 33.1

Estimate 44 + 33 = _____

1 Line up the decimal points.

2 Add the digits in the same place-value positions.

3 Bring the decimal point straight down in the sum.

$$
\begin{array}{r}
4\ \ 4\ .\ 2 \\
+\ 3\ \ 3\ .\ 1 \\
\hline
\boxed{}\ \boxed{}\ .\ \boxed{}
\end{array}
$$

So, the total amount of sodium used is _____ milligrams.

Check for Reasonableness

Compare to the estimate, _____ .

_____ ≈ _____

	Tens	Ones	Tenths
	4	4	2
+	3	3	1
	7	7	3

You can annex a zero to one of the numbers in an addition problem so that both numbers end in the same place value.

Example 2

Find 19.6 + 4.31.

Estimate 20 + 4 = _____

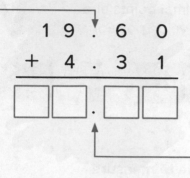

1 Line up the decimal points. Annex a 0 so that both numbers end in the same place value.

$$
\begin{array}{r}
1\ 9\ .\ 6\ 0 \\
+\ \ 4\ .\ 3\ 1 \\
\hline
\square\ \square\ .\ \square\ \square
\end{array}
$$

2 Add the digits in the same place-value positions. Rename as necessary.

3 Bring the decimal point straight down in the sum.

So, 19.6 + 4.31 = _____.

Check for Reasonableness Compare to the estimate, _____.

_____ ≈ _____

Guided Practice

Add. Check for reasonableness.

1.
$$
\begin{array}{r}
6\ .\ 3\ 2 \\
+\ 1\ .\ 4\ 6 \\
\hline
\square\ .\ \square\ \square
\end{array}
$$

Estimate 6 + 1 = _____

So, 6.32 + 1.46 = _____.

Check for Reasonableness

_____ ≈ _____

2.
$$
\begin{array}{r}
0\ .\ 8\ 9 \\
+\ 0\ .\ 0\ 3 \\
\hline
\square\ .\ \square\ \square
\end{array}
$$

Estimate 1 + 0 = _____

So, 0.89 + 0.03 = _____.

Check for Reasonableness

_____ ≈ _____

Talk MATH

Explain how annexing zeros might be helpful when adding decimals.

Independent Practice

Add. Check for reasonableness.

3.   0.54
        $+7.8$

4.   14.8
        $+10.26$

5.   25
        $+8.46$

6.   35.08
        $+11.9$

7.   0.8
        $+0.22$

8.   9.14
        $+2.05$

9. $6.57 + 1.2 =$ _____

10. $19.21 + 11.03 =$ _____

11. $3.08 + 1.64 =$ _____

Algebra Find each unknown.

12. $8.9 + 0.15 = x$

$x =$ _____

13. $42.2 + 7.69 = d$

$d =$ _____

14. $5.63 + 1.22 = w$

$w =$ _____

Copyright © McGraw-Hill Education

Lesson 6 Add Decimals   **337**

Problem Solving

15. Horacio bought a logic puzzle and batteries from a toy store. Use the table at the right to find the total cost of the two items, not including tax.

Item	Cost ($)
logic puzzle	14.95
batteries	10.39
carrying case	12.73

_____

16. An athlete training for the Olympics swims each lap of a four-lap race in the following times: 54.73, 54.56, 54.32, and 54.54 seconds. What is the total time it takes her to swim the four laps?

_____

17. Terrance is bicycling on a trail. He bicycles for 12.6 miles and takes a break. Then he bicycles for 10.7 miles. How many miles has Terrance bicycled in all?

_____

Brain Builders

18. **Processes &Practices** **1** **Make a Plan** Write a real-world word problem that can be solved by adding 1.99 and 2.49. Then draw a model to represent the solution.

_____

_____

19. **Processes &Practices** **2** **Use Number Sense** Write two different pairs of decimals whose sums are 8.69. One pair should involve regrouping.

_____

20. **Building on the Essential Question** How does place value help you add decimals? Show an example.

_____

_____

Name ....

MY Homework

Homework Helper

Need help? 🔎 **connectED.mcgraw-hill.com**

Brett's social studies book weighs **5.34 pounds**. His science book weighs **4.78 pounds**. Suppose Brett only has these books in his bookbag. How much weight is he carrying, not including the weight of his bookbag?

Find $5.34 + 4.78$.

Estimate $5 + 5 = 10$

1 Line up the decimal points.

2 Add the digits in the same place-value positions. Rename as necessary.

$$
\begin{array}{r}
\overset{1}{5}.\overset{1}{3}4 \\
+\ 4.78 \\
\hline
10.12
\end{array}
$$

3 Bring the decimal point straight down in the sum.

So, Brett is carrying 10.12 pounds in books.

Check for Reasonableness $10.12 \approx 10$

Practice

Add. Check for reasonableness.

1. $\begin{array}{r} \$2.72 \\ +\ \$3.83 \\ \hline \end{array}$

2. $\begin{array}{r} 12.03 \\ +\ 0.14 \\ \hline \end{array}$

3. $\begin{array}{r} 26.76 \\ +\ 2.99 \\ \hline \end{array}$

![Real World] **Problem Solving**

4. Devin has a new cell phone that holds 1.5 gigabytes of memory for music. He has already used 1.35 gigabytes of the memory. Will he have enough space to download a song that will use 0.12 gigabyte of memory? Explain.

5. A large bag of sand weighs 48.5 pounds. A small bag of sand weighs 24.6 pounds. If Mrs. Waggoner buys a large bag and a small bag, how many pounds of sand did she purchase altogether?

Brain Builders

6. Processes &Practices **6** **Explain to a Friend** Grady wants to buy a basketball video game that costs $59.95, plus $2.39 in tax. He has $45.50 in cash and a gift certificate for $15.25. Is that enough to buy the video game? Explain to a friend.

7. Processes &Practices **7** **Identify Structure** Lakshmi wants to start saving coins in a piggy bank. Her mother gave her three quarters and two pennies on Monday and two dimes and one nickel on Tuesday. Fill in the amounts of money that Lakshmi's mother gave her in the space provided on the bar diagram below.

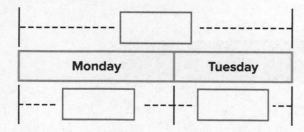

8. Test Practice Marcus entered a race that involves swimming and running. He will need to swim 0.72 mile and run 1.65 miles. How far will Marcus travel in all during the race?

Ⓐ 2.37 miles   Ⓒ 2.07 miles

Ⓑ 2.17 miles   Ⓓ 1.37 miles

Lesson 7
Addition Properties

ESSENTIAL QUESTION ❓
How can I use place value and properties to add and subtract decimals?

You can use properties of addition to find sums of whole numbers and decimals mentally. When there are no parentheses, add in order from left to right.

 Math in My World Watch ▶ Tutor 💬

Example 1

Elijah recorded the number of different movies he watched last month. Use properties of addition to mentally find the total number of movies.

Find 5 + 27 + 15.

You can easily add 5 and 15. So, change the order and group those numbers together.

Movie Genre	Number of Movies Watched
Comedy	5
Action	27
Drama	15

The **Commutative Property of Addition** states that the order in which numbers are added does not change the sum.

The **Associative Property of Addition** states that the way in which numbers are grouped does not change the sum.

5 + 27 + 15 = _____ + 5 + 15   Commutative Property

= 27 + (_____ + _____)   Associative Property

= 27 + _____   Add mentally. 5 + 15 = _____

= _____   Add mentally. 20 + 27 = _____

So, Elijah watched a total of _____ movies.

The **Identity Property of Addition** states that the sum of any number and 0 equals the number.

Example 2

Use properties to find 1.8 + 2.6 + 0 mentally.

$$1.8 + 2.6 + 0 = (\underline{\hspace{1cm}} + 0.8) + (\underline{\hspace{1cm}} + 0.6) + 0$$

$= 1 + \underline{\hspace{1cm}} + 0.8 + \underline{\hspace{1cm}} + 0$	Commutative Property
$= (1 + \underline{\hspace{1cm}}) + (0.8 + \underline{\hspace{1cm}}) + 0$	Associative Property
$= \underline{\hspace{1cm}} + \underline{\hspace{1cm}} + 0$	Add.
$= \underline{\hspace{1cm}} + 0$	Identity Property
$= \underline{\hspace{1cm}}$	Add.

Guided Practice

1. Use properties of addition to find each sum mentally. Show your steps and identify the properties that you used.

$9 + 27 + 1 = \underline{\hspace{1cm}} + 9 + 1$	Commutative Property
$= 27 + (\underline{\hspace{1cm}} + \underline{\hspace{1cm}})$	Associative Property
$= 27 + \underline{\hspace{1cm}}$	Add.
$= \underline{\hspace{1cm}}$	Add.

Use properties to mentally determine whether 3.1 + 0.8 + 0.9 is less than, greater than, or equal to 5. Explain.

Independent Practice

Identify the properties that are used to find each sum.

2. $69 + 22 = (60 + 9) + (20 + 2)$

$= 60 + 20 + 9 + 2$ _____

$= (60 + 20) + (9 + 2)$ _____

$= 80 + 11$ Add.

$= 91$ Add.

3. $11 + 7.7 + 4.3 + 0 = 11 + (7.7 + 4.3) + 0$ _____

$= 11 + 12 + 0$ Add.

$= 23 + 0$ _____

$= 23$ Add.

4. $37 + 26 + 53 = 26 + 37 + 53$ _____

$= 26 + (37 + 53)$ _____

$= 26 + 90$ Add.

$= 116$ Add.

5. Use properties of addition to find the sum mentally. Show your steps and identify the properties that you used.

$10.9 + 3 + 0.1 =$ _____

Processes
6. &Practices  **2 Use Number Sense** Casey spent $2.50 on a snack, $1.24 on gum, $3.76 on a comic book, and $5.50 on lunch. Use mental math to find the total amount that he spent.

_____

7. In one week, a classroom collected 43, 58, 62, 57, and 42 cans. Find the total number of cans the classroom collected using mental math. Explain how you found the sum.

_____

_____

Brain Builders

Cost of a Cheerleading Uniform	
Shoes	$65
Pom-poms	$18
T-shirt and shorts	$35

8. The table shows the cost of a cheerleading uniform. Explain how you can use the properties of addition to find the total cost of the uniform mentally. Then solve.

_____

_____

Processes
9. &Practices 4 Model Math Write a real-world problem that can be solved using the Associative Property of Addition. Solve the problem. Write a number model to show how you found the sum.

_____

_____

_____

_____

10. ? Building on the Essential Question How can properties help me add whole numbers and decimals? Give an example.

_____

_____

_____

Name ....

Homework Helper

Need help? ☞ **connectED.mcgraw-hill.com**

Mackenzie practices violin four days a week. One week, she practiced for 21, 39, 45, and 25 minutes. Use mental math to find the total amount of time she practiced.

You can easily add 21 and 39. You can easily add 45 and 25. So, group those numbers together.

$21 + 39 + 45 + 25 = (21 + 39) + (45 + 25)$ Associative Property
$= 60 + 70$ Add. $21 + 39 = 60$ and $45 + 25 = 70$
$= 130$ Add. $60 + 70 = 130$

So, Mackenzie practiced violin for 130 minutes.

Practice

Use properties of addition to find each sum mentally. Show your steps and identify the properties that you used.

1. $9 + 6 + 31 =$ _____

2. $12.5 + 0 + 1 + 43.5 =$ _____

3. Sasha spent $1.05 on a soda, $5.25 on a sandwich, $0.75 on a piece of fruit, and $4.95 on a magazine. Use mental math to find the total amount she spent.

_____

4. Jessie went to the mall and bought a CD for $12.98, a skirt for $17.50, a T-shirt for $8.50, and a bottle of water for $1.02. Use mental math to find the total amount she spent.

_____

Brain Builders

5. Gary played soccer for 1 hour, volleyball for 3.5 hours, and tennis for 2 hours. Tanya played soccer for 3.5 hours, volleyball for 2 hours, and tennis for 1 hour. Who played sports longer? Explain.

_____

_____

_____

6. **Processes &Practices** **5** **Use Math Tools** Without calculating, would $0.4 + (2 + 0.6)$ be less than, greater than, or equal to 3? Which property or properties would you use to determine the answer? Explain.

_____

_____

_____

7. **Test Practice** Paula was reading a novel. She read 13 pages on Sunday, 12 pages on Tuesday, 17 pages on Friday, and 8 pages on Saturday. Use mental math to find the total number of pages she read.

Ⓐ 40 pages   Ⓒ 50 pages

Ⓑ 42 pages   Ⓓ 60 pages

Check My Progress

Vocabulary Check

Match each word to its correct definition.

1. **Commutative Property of Addition**

2. **Associative Property of Addition**

3. **Identity Property of Addition**

- The sum of any number and 0 equals the number.

- The way in which numbers are grouped does not change the sum.

- The order in which numbers are added does not change the sum.

Concept Check

Add.

4. $3.15 + 1.2 =$ _____

5. $68.9 + 7.1 =$ _____

6. $\begin{array}{r} 4.67 \\ + 1.70 \\ \hline \end{array}$

7. $\begin{array}{r} 25.39 \\ + 18.68 \\ \hline \end{array}$

8. Use properties of addition to find $37 + 58$ mentally.
Show your steps and identify the properties that you used.

$37 + 58 =$ _____

GLACIER UNIVERSITY

$25.15

$19.74

9. What is the combined cost of the sweatshirt and hat shown?

_____

10. Francis earned the following scores on his math quizzes: 86, 84, 90. Jillian earned the following scores on her math quizzes: 84, 90, 86. Who earned a greater total score? Which addition property can you use to find the answer?

_____

_____

Brain Builders

11. Dean received the following amounts for mowing his neighbors' lawns: $12.34, $16.75, $20.66, $33.25. Use mental math to find the total amount he earned.

_____

_____

12. Explain how you would find the sum of 4.2 and 2.14.

_____

_____

_____

13. Test Practice Tammy made a bracelet using red, white, and blue string. The red string is 2.4 centimeters long, the white string is 2.1 centimeters long, and the blue string is 2.6 centimeters long. What is the total length of the three strings?

Ⓐ 7.1 centimeters   Ⓒ 5.5 centimeters

Ⓑ 6.1 centimeters   Ⓓ 4.27 centimeters

Lesson 8
Hands On
Subtract Decimals Using Base-Ten Blocks

ESSENTIAL QUESTION
How can I use place value and properties to add and subtract decimals?

Build It Tools

Find 1.8 — 0.4. Use base-ten blocks.

1 Model 1.8.

Ones	Tenths	Hundredths

2 Take 0.4, or four tenths, away.

Count the remaining base-ten blocks. Draw the result.

How many ones are left? _____

How many tenths are left? _____

So, 1.8 − 0.4 = _____.

Check Use addition to check your answer.

_____ + 0.4 = 1.8

Find 2.25 − 0.75. Use base-ten blocks.

1 Model 2.25.

2 Subtract 0.75. Count the remaining base-ten blocks. Draw the result.

To take away 0.75, take away 7 tenths and 5 hundredths. But you cannot subtract 7 tenths from 2 tenths. So, regroup the ones block as 10 tenths. Then subtract.

Ones	Tenths	Hundredths

How many ones are left? _____

How many tenths are left? _____

How many hundredths are left? _____

So, 2.25 − 0.75 = _____ .

Check Use addition to check your answer.

_____ + 0.75 = 2.25

Helpful Hint

2 ones − 1 one = 1 one

12 tenths − 7 tenths = 5 tenths

5 hundredths − 5 hundredths = 0 hundredths

Talk About It

1. **Processes & Practices** **6** **Explain to a Friend** Explain to a friend or classmate when you should regroup when subtracting decimals with base-ten blocks.

_____

_____

_____

2. How many tenths blocks will you need to subtract 0.35 from 1.2? Explain.

_____

Practice It

Subtract. Use base-ten blocks. Draw each result in the table.

3. $0.8 - 0.3 =$ _____

Ones	Tenths	Hundredths

4. $2.8 - 0.7 =$ _____

Ones	Tenths	Hundredths

5. $1.43 - 0.31 =$ _____

Ones	Tenths	Hundredths

6. $2.17 - 1.9 =$ _____

Ones	Tenths	Hundredths

7. $2.52 - 1.48 =$ _____

Ones	Tenths	Hundredths

8. $3.85 - 2.19 =$ _____

Ones	Tenths	Hundredths

Apply It

9. Vaughn's MP3 player had a memory of 2.5 gigabytes. He has used 1.76 gigabytes of memory. How much memory space does Vaughn's MP3 player have left? Use base-ten blocks to help you solve.

_____

10. Yolanda cut a piece of wood that was 1.6 feet long from a 3.25-foot piece of wood. How long is the remaining piece of wood? Use base-ten blocks to help you solve.

_____

11. **Processes &Practices** [4] **Model Math** Write a real-world problem that can be solved using the base-ten blocks below.

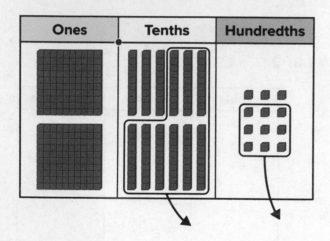

_____

_____

Write About It

12. How can I use base-ten blocks to subtract decimals?

_____

_____

_____

Name _____

Homework Helper

Need help? ↗ connectED.mcgraw-hill.com

Charlotte bicycled 1.6 miles on Tuesday and 0.9 mile on Wednesday. How much farther did she bike on Tuesday than on Wednesday?

Find $1.6 - 0.9$. Use base-ten blocks.

1 Model 1.6.

Ones	Tenths	Hundredths

2 Subtract 0.9. Count the remaining base-ten blocks.

Ones	Tenths	Hundredths

You need to take away 9 tenths, but you cannot subtract
9 tenths from 6 tenths. So, regroup the ones block as
10 tenths. Then subtract.
There are no ones left.
There are seven tenths left.

So, $1.6 - 0.9 = 0.7$.

Charlotte biked 0.7 mile farther on Tuesday.

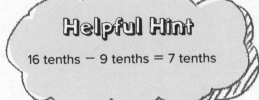

Helpful Hint

16 tenths − 9 tenths = 7 tenths

Check Use addition to check your answer.

$0.7 + 0.9 = 1.6$

Practice

Subtract. Use base-ten blocks. Draw each result in the table.

1. 1.3 − 0.28 = _____

Ones	Tenths	Hundredths

2. 3.52 − 1.39 = _____

Ones	Tenths	Hundredths

Problem Solving

3. **Processes &Practices 4** **Model Math** The temperature at 6:00 A.M. was 0.8°F. By 3:00 P.M., the temperature had increased to 2.4°F. Find the difference between the two temperatures. Draw models to find the difference.

_____

4. Melissa calculated that she can swim 3.6 miles per hour. Avery calculated her swimming speed to be 2.8 miles per hour. How much faster was Melissa's swimming speed? Draw models to find the difference.

_____

5. Monica paid for a milkshake using $3. If the milkshake costs $2.67, how much change will she receive? Draw models to find the difference.

_____

Name _____

ESSENTIAL QUESTION
How can I use place value and properties to add and subtract decimals?

Build It Tools

Find 2.4 − 1.07. Use models.

1 **Use 10-by-10 grids to model 2.4.**
To show 2.4, shade two whole grids and four tenths of a third grid.

Four tenths is equal to _____ small squares.

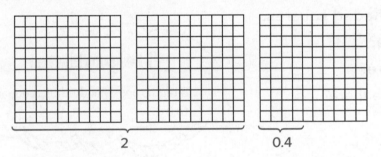

2   0.4

2 **Subtract 1.07.**
To subtract 1.07, cross out 1 whole grid and 7 hundredths of the third grid.

3 **Count the remaining shaded squares.**

How many ones are there? _____

How many tenths are there? _____

How many hundredths are there? _____

So, 2.4 − 1.07 = _____.

Check Use addition to check your answer.

1.07 + _____ = 2.4

Find 1.66 − 0.84. Use models.

 Model 1.66.

To show 1.66, shade one grid and sixty-six hundredths of a second grid.

Sixty-six hundredths is equal to _____ small squares.

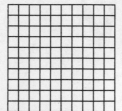 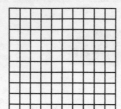

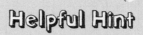 **Subtract 0.84.**

To subtract 0.84, cross out 4 hundredths and 8 tenths.

3 Count the remaining shaded squares.

How many ones are there? _____

How many tenths are there? _____

How many hundredths are there? _____

So, 1.66 − 0.84 = _____ .

Check Use addition to check your answer.

0.84 + _____ = 1.66

> **Helpful Hint**
>
> 8 tenths and 2 hundredths is equal to 0.82.

Talk About It

Processes
1. **&Practices** **5** **Use Math Tools** Explain how using models to find
2.4 − 1.07 is similar to using models to find 240 − 107.

_____

_____

_____

_____

Practice It

Subtract. Use decimal models.

2. $0.93 - 0.7 =$ _____

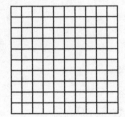

3. $1.53 - 1.41 =$ _____

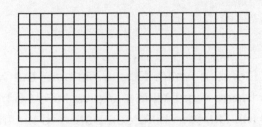

4. $0.9 - 0.3 =$ _____

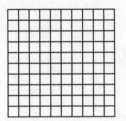

5. $3.94 - 0.4 =$ _____

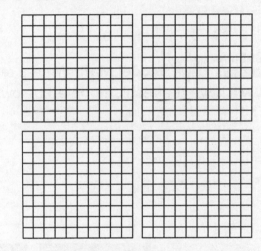

6. $3.55 - 0.1 =$ _____

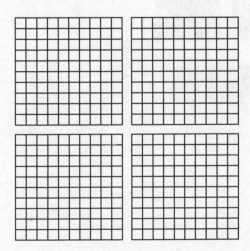

7. $2.4 - 0.9 =$ _____

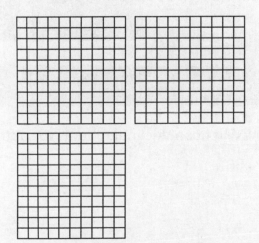

Processes &Practices **5** **Use Math Tools** Nora wrote two papers for English class. One uses 0.65 megabyte of memory and the other uses 0.92 megabyte of memory. How much more memory does the one paper use? Use decimal models to help you solve.

_____

9. Mr. Dorsten walked the trails over the weekend. He hiked 2.3 miles. His friend went hiking and walked 1.6 miles. How many more miles did Mr. Dorsten hike compared to his friend? Use decimal models to help you solve.

_____

10. **Processes &Practices** **4** **Model Math** Write a real-world subtraction problem that will have the result modeled below. Then solve.

_____

_____

_____

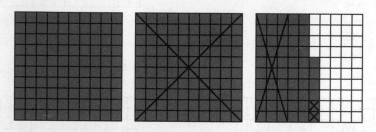

Write About It

11. How can I use decimal models to subtract decimals?

_____

_____

_____

MY Homework

Homework Helper

Need help? connectED.mcgraw-hill.com

Marcy measured the lengths of two insects. The first insect was 2.43 centimeters long. The second insect was 1.05 centimeters long. What is the difference between the lengths of the two insects?

Find 2.43 − 1.05.

1 **Use 10-by-10 grids to model 2.43.**

To show 2.43, shade two grids and forty-three hundredths of a grid.

Forty-three hundredths is equal to 43 small squares.

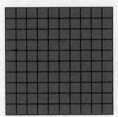 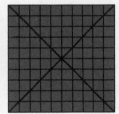 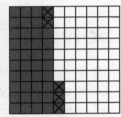

2 **Subtract 1.05.**

To subtract 1.05, cross out 1 whole grid and 5 hundredths of the third grid.

3 **Count the remaining shaded squares.**

There is 1 one.

There are 3 tenths.

There are 8 hundredths.

Helpful Hint

1 one, 3 tenths, and 8 hundredths is equal to 1.38.

So, 2.43 − 1.05 = 1.38.

The difference in the lengths of the two insects is 1.38 centimeters.

Check Use addition to check your answer.

1.05 + 1.38 = 2.43

Practice

Subtract. Use decimal models.

1. 3.8 − 2.3 = _____

2. 2.13 − 1.7 = _____

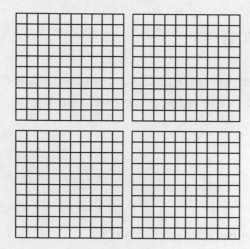

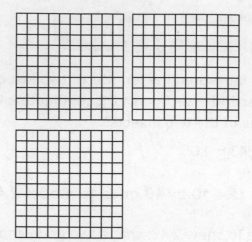

 Problem Solving

3. Bridget is reviewing a nutrition label. She wants to find the difference between the amount of grams of saturated fat and protein. How many more grams are there of saturated fat than protein? Draw decimal models to find the difference.

_____

Nutrition Facts		
Serving Size: 1 tbsp (15 g)		
Amount Per Serving		
Calories 190	**Calories from Fat 1**	
		% Daily Value*
Total Fat 1.48 g		0%
Saturated Fat 1.08 g		0%
Trans Fat		
Cholesterol 0 mg		0%
Sodium 164.4 mg		7%
Potassium 1.4 mg		0%
Total Carbohydrate 5.17 g		2%
Dietary Fiber 0.48 g		2%
Sugars		
Sugar Alcohols		
Protein 0.49 g		

Processes & Practices **5** **Use Math Tools** Leanne is building a model tower that needs to be 1.42 meters tall. The model currently is only 0.68 meter tall. How many more meters does she need to build in order to finish the model tower? Draw decimal models to find the difference.

_____

Name _____

Lesson 10
Subtract Decimals

ESSENTIAL QUESTION
How can I use place value and properties to add and subtract decimals?

To subtract decimals, line up the decimal points. Then subtract digits in the same place-value position.

 Math in My World Watch ▶ Tutor 💬

Example 1

The table shows the average lengths of the three longest bones in the human body. How much longer is the average femur than the average tibia?

Find 19.8 − 16.9.

| Longest Bones in the Human Body ||
Bone	Length (in.)
Femur (upper leg)	19.8
Tibia (inner lower leg)	16.9
Fibula (outer lower leg)	15.9

Estimate 20 − 17 = _____

 Line up the decimal points. ⌐─────────────────┐

$$
\begin{array}{r}
\square\ \ \square \\
-\ 1\ 9\ .\ 8 \\
-\ 1\ 6\ .\ 9 \\
\hline
\square\ .\ \square
\end{array}
$$

 Subtract the digits in the same place-value positions. Rename as necessary.

 Bring the decimal point straight down in the difference. ⌐───────────────┘

So, the average femur is _____ inches longer than the average tibia.

Subtraction and addition are inverse operations. **Inverse operations** are operations that undo each other.

Check Use addition to check your answer.

_____ + 16.9 = 19.8.

Sometimes both numbers in a subtraction problem do not end in the same place value. Annex zeros if necessary.

Example 2

Find 6.3 − 4.78.

Estimate 6 − 5 = _____

1. Line up the decimal points. Annex a 0 so that both numbers end in the same place value.

2. Subtract the digits in the same place-value positions. Rename as necessary.

3. Bring the decimal point straight down in the difference.

```
    ☐ . ☐   ☐
    6 . 3   0
  − 4 . 7   8
    ☐ . ☐   ☐
```

So, 6.3 − 4.78 = _____.

Check Use addition to check your answer.

_____ + 4.78 = 6.3

Guided Practice

Subtract. Use addition to check your answer.

1.
```
    5 . 5
  − 3 . 2
    ☐ . ☐
```

Estimate 6 − 3 = _____

So, 5.5 − 3.2 = _____.

Check

_____ + 3.2 = 5.5

2.
```
    ☐  ☐ . ☐
    7  2 . 4
  − 1  2 . 5
    ☐  ☐ . ☐
```

Estimate 70 − 10 = _____

So, 72.4 − 12.5 = _____.

Check

_____ + 12.5 = 72.4

Explain the strategy used in Exercise 1 to subtract the decimals.

362 Chapter 5 Add and Subtract Decimals

Independent Practice

Subtract. Use addition to check your answer.

3. 29.34
     − 9

4. 0.4
     − 0.2

5. $9.67
     − $2.35

6. 97
     − 16.98

7. 42.28
     − 1.52

8. 8
     − 5.78

9. 36 − 7.3 = _____

10. 5.6 − 3.5 = _____

11. 19.86 − 9.94 = _____

Algebra Find each unknown.

12. $15.00 − $6.24 = b

b = _____

13. 8.2 − 6.72 = k

k = _____

14. 58.67 − 28.72 = t

t = _____

Problem Solving

15. Use the table to find out how many more people there are per square mile in Iowa than in Colorado.

_____

Population Density	
State	People Per Square Mile
Colorado	41.5
Iowa	52.4

16. **Processes &Practices** **6** **Explain to a Friend** You decide to buy a hat for $10.95 and a T-shirt for $14.20. How much change will you receive if you pay with a $50 bill? Explain to a friend.

_____

_____

Brain Builders

17. Coach Trainer timed his athletes running the 40-yard dash. The fastest time was 4.58 seconds and the difference between the slowest and fastest times was 1.17 seconds. Find the slowest time.

_____

_____

18. **Processes &Practices** **2** **Reason** Explain how place value could be used to find the difference of 4.23 and 2.75.

_____

_____

19. **?** **Building on the Essential Question** How does place value help me subtract decimals?

_____

_____

_____

MY Homework

Homework Helper

Need help? connectED.mcgraw-hill.com

Stephen's father gave him $10 to buy lunch at the concession stand. If his lunch cost $7.74, how much change should Stephen give his father?

Find $10 − $7.74.

Estimate $10 − 8 = 2$

1. Line up the decimal points. Annex zeros so that both numbers end in the same place value.

2. Subtract the digits in the same place-value positions. Rename as necessary.

3. Bring the decimal point straight down in the difference.

$$
\begin{array}{r}
0 \quad 9 \quad 9 \quad 10 \\
\$\,\cancel{1}\,\cancel{0}\,.\,\cancel{0}\,\cancel{0} \\
-\ \$\quad 7\,.\,7\ 4 \\
\hline
\$\quad 2\,.\,2\ 6
\end{array}
$$

So, Stephen should give his father $2.26.

Check Use addition to check your answer.

$2.26 + $7.74 = $10.00

Practice

Subtract. Use addition to check your answer.

1. 9.03
   − 4.09

2. 71.65 − 55.12 = _____

3. $18. 9 4
   − $3. 2 5

4. Roberto uses his handheld GPS device to determine that he hikes 21.48 miles in one weekend. The next weekend he hikes 30 miles. How much less did he hike the first weekend than the second weekend?

_____

5. Jabir buys 2.74 pounds of dried pinto beans and 4.05 pounds of dried lima beans. What is the difference between the weights of the beans that Jabir buys?

_____

Brain Builders

6. Rebekah's baby brother weighs 7.71 pounds. How much more than 10 pounds will her baby brother weigh if he gains 1.3 pounds each week for the next 2 weeks?

_____

7. **Processes &Practices** 2 **Use Number Sense** Marshall has been given the option to receive an allowance once a week or once a month. Should he choose $3.25 each week or $10.75 each month? Explain.

_____

_____

_____

8. **Test Practice** Kamal buys a pen that costs $1.09 and a tablet of paper that costs $2.50. How much more does the paper cost than the pen?

Ⓐ $1.31   ⓒ $1.59

Ⓑ $1.41   Ⓓ $3.59

Vocabulary Check

Use the words in the word bank to complete each sentence.

Associative Property	Commutative Property	Identity Property
inverse operations	estimate	difference
exact answer	place value	regroup

1. The _____ states that you can add numbers in any order.

2. When you round a number, you find its _____.

3. The value given to a digit by its position in a number is its

   _____.

4. To _____ is to form into a new or restructured group or grouping.

5. Subtracting results in finding the _____ between two numbers.

6. The _____ states that the sum of any number and 0 equals the number.

7. The _____ states that the way in which numbers are grouped does not change the sum or product.

8. When computing using actual numbers, you find a(n)

   _____.

Concept Check

Round each decimal to the place indicated.

9. 0.781; hundredths _____

10. 17.93; ones _____

Round each decimal to the nearest tenth. Then add or subtract.

11. 17.16 − 9.01 = _____

12. 62.84 + 74.2 = _____

Add.

13.
```
   6.07
+  1.34
```

14.
```
   $1.06
+  $4.89
```

15. 4.15 + 20.68 = _____

16. Use properties of addition to find 37 + 48 mentally. Show your steps and identify the properties that you used.

37 + 48 = _____

Subtract. Use addition to check your answer.

17.
```
   4.15
−  1.29
```

18.
```
   $13.78
−  $5.41
```

19. 8.3 − 4.7 = _____

Problem Solving

For Exercises 20 and 21, determine whether you need an estimate or an exact answer. Then solve.

20. A family is staying in a hotel for 2 nights. The hotel costs $91.35 a night. About how much will they pay for the hotel?

_____

_____

21. Jesse had three long jumps during a track meet of 17 feet, 18 feet, and 15 feet. Andrew had three long jumps during a track meet of 17 feet, 15 feet, and 18 feet. Overall, who had a greater total jump distance? Which addition property can you use to find the answer?

_____

_____

_____

Brain Builders

22. You have to be 48 inches tall to ride the roller coaster. Eva is 45 inches tall. If she grows 1.5 inches each year, will she be able to ride the roller coaster in 2 years? Write a number model to support your answer.

_____

_____

23. Test Practice Micah had $10.52 after he left the book store. If he bought a book for $6.39, and a magazine for $3.99, how much money did he have before he went to the book store?

Ⓐ $0.14   Ⓒ $16.91

Ⓑ $10.38   Ⓓ $20.90

Reflect

Use what you learned about adding and subtracting decimals to complete the graphic organizer.

Write the Example

Real-World Example

ESSENTIAL QUESTION

How can I use place value and properties to add and subtract decimals?

Vocabulary

Models

Now reflect on the **ESSENTIAL QUESTION** Write your answer below.

_____

_____

Performance Task

Planning for a Mountain Bike Trip

Sarah and her two friends are planning for a mountain bike trip. Each of them will donate $68 to the park. The park uses donations to maintain the trails and facilities. They will follow three different trails, and the table shows the total elevation of each trail.

Red Trail	364.78 feet
Yellow Trail	645.42 feet
Green Trail	438.36 feet

Show all your work to receive full credit.

Part A

Estimate the total number of feet that Sarah and her friends will climb during the mountain bike trip by rounding the elevation for each trail to the nearest 10. Show your estimates.

Part B

Every 75 feet, they will drink 8 ounces of water. Use your estimate to determine how many times they will stop to drink 8 ounces of water. How many ounces of water will Sarah drink biking up the mountain?

Part C

Sarah has planned the following expenses for the group for meals and lodging at the inn located at the top of the mountain.

Breakfast	$14.97
Lunch	$21.75
Dinner	$38.85
Inn	$152.00

The first night the group will eat dinner and stay at the inn. The next day they eat all three meals and stay at the inn. The next morning they eat breakfast and start biking down the mountain. How much should the group expect to spend on food and lodging? Explain.

Part D

Sarah and her friends saved $840 altogether for the trip. Given the donation each person will give to the park and the expenses for meals and lodging, how much can they expect to have left at the end of the mountain bike trip? Explain.

ESSENTIAL QUESTION

How is multiplying and dividing decimals similar to multiplying and dividing whole numbers?

My Summer Fun

Watch

Watch a video!

MY Chapter Project

Tree House

1. Create a plan for your dream tree house. Using a blank sheet of paper, sketch what you want it to look like, and list the materials you would use to build it, such as wood, nails, paint, and decorations.

2. After your design is complete, use the Internet to find the costs of all the building materials you listed. Write the prices next to the items on your list.

3. Use multiplication of decimals to find the total cost of building the tree house.

4. After you have finished, wait for your teacher to put you and your classmates into groups.

5. Determine the cost per person if you split the cost of your tree house with the other members in your group. Do this for the tree house costs of each of your group members.

6. Finally, present your tree house proposal to your class, explaining the materials you used and how much the whole house would cost. Each time a classmate presents, give their proposal a score on a scale of 1 to 10.

Am I Ready?

Multiply.

1. 126 × 30 = _____

2. 12 × 28 = _____

3. 320 × 10 = _____

4. How much will it cost for Gary to reserve 13 lanes?

Bowling Reservations	
1 lane	$12

_____

Divide.

5. 114 ÷ 6 = _____

6. 84 ÷ 4 = _____

7. 1,500 ÷ 10 = _____

8. Hugo spent $216 on 4 sweaters. If each sweater cost the same amount, find the cost of one sweater.

_____

Round each decimal to the nearest whole number.

9. 2.7 _____

10. 0.7 _____

11. 18.2 _____

12. 6.34 _____

13. 9.8 _____

14. 9.4 _____

How Did I Do? Shade the boxes to show the problems you answered correctly.

| 1 | 2 | 3 | 4 | 5 | 6 | 7 | 8 | 9 | 10 | 11 | 12 | 13 | 14 |

Online Content at connectED.mcgraw-hill.com

MYMath Words Vocab abc

Review Vocabulary

composite number   decimal point   divide   estimate

hundredths   multiply   ones   place value

power   tenths   thousands

Making Connections

Use the Venn diagram to categorize the review vocabulary.

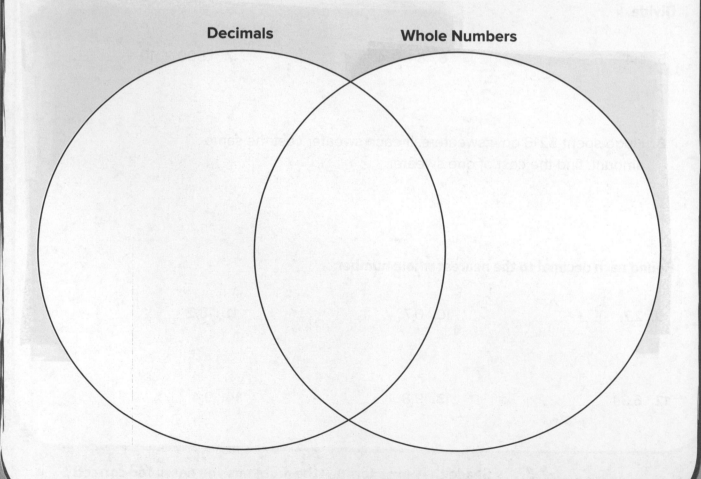

Decimals   Whole Numbers

MY Vocabulary Cards

Processes
&Practices

Lesson 6-8

Associative Property of Multiplication

$$(8 \times 7) \times 10 = 8 \times (7 \times 10)$$

Lesson 6-8

Commutative Property of Multiplication

$$8.92 \times 455 = 455 \times 8.92$$

Lesson 6-8

Identity Property of Multiplication

$$42.08 \times 1 = 42.08$$

Copyright © McGraw-Hill Education

Ideas for Use

- Use blank cards to group like ideas you find throughout the chapter, such as multiplying using decimals and dividing using decimals. Discuss with a friend strategies to understand these concepts.

- Use a blank card to write this chapter's essential question. Use the back of the card to write examples that help you answer the question.

The order in which factors are multiplied does not change the product.

How is the Commutative Property of Multiplication different from the Identity Property of Multiplication?

_____

_____

The way in which factors are grouped does not change the product.

How can the word associate help you remember this property?

_____

_____

The product of any factor and 1 equals the factor.

Write a number sentence that is an example of the Identity Property of Multiplication.

_____

MY Foldable

FOLDABLES® Follow the steps on the back to make your Foldable.

✂

Power of 10

exponent   power of 10

5.1

5.1×10^2 ▶ **510** ◀ 5.1×100

5.1×10^3 ▶ **5,100** ◀ $5.1 \times 1,000$

5.1×10^1 ▶ **51** ◀ 5.1×10

There is 1 zero in 10.
So, the exponent is 1.
$$10 = 10^1$$

There are 2 zeros in 100.
So, the exponent is 2.
$$100 = 10^2$$

There are 3 zeros in 1,000.
So, the exponent is 3.
$$1,000 = 10^3$$

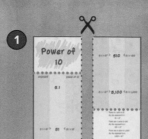

① Power of 10

②

③ Power of 10

Lesson 1
Estimate Products of Whole Numbers and Decimals

ESSENTIAL QUESTION ?

How is multiplying and dividing decimals similar to multiplying and dividing whole numbers?

Math in My World

  Watch Tutor

Example 1

Mrs. James buys about 15.8 gallons of gas each week for her car. About how many gallons of gas will she buy in 5 weeks?

Estimate the product of 15.8 and 5.

 Round.

$$15.8 \longrightarrow \boxed{}$$
$$\underline{\times\ 5} \qquad\qquad \underline{\times\ 5}$$

 Then multiply to estimate the product. $\longrightarrow \boxed{}$

So, 15.8×5 is about _____ .

Mrs. James buys about _____ gallons of gas in 5 weeks.

Helpful Hint

Since 15.8 rounded up to 16, the estimated product is greater than the actual product.

Check Use repeated addition.

$16 + 16 + 16 + 16 + 16 =$ _____ or

$15.8 + 15.8 + 15.8 + 15.8 + 15.8 =$ _____

_____ is close to the estimate, _____ , so the answer is reasonable.

Example 2

A worker loads 16 packages onto a truck. Each package weighs 58.5 pounds. About how many pounds are loaded onto the truck?

Estimate the product of 58.5 and 16.

 Use compatible numbers.

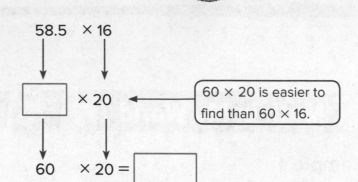

$58.5 \quad \times 16$

$\boxed{} \quad \times 20$ ← 60 × 20 is easier to find than 60 × 16.

 Multiply.

$60 \quad \times 20 = \boxed{}$

So, 58.5 × 16 is about _____.

Guided Practice

Estimate each product.

1. 2 × $4.10

_____

2. 18.4 × 10

_____

Explain how to round 18.9 to the nearest whole number.

3. 19.8 × 3

_____

4. 5 × $2.14

_____

Independent Practice

Estimate each product.

5. 4 × $4.62

_____

6. 3 × $23.07

_____

7. $15.50 × 6

_____

8. $16.85 × 9

_____

9. 7.2 × 5

_____

10. 14.5 × 3

_____

11. 9 × 19.7

_____

12. 11 × 26.2

_____

13. $0.89 × 14

_____

14. 18.8 × 13

_____

Problem Solving

Use the items below to estimate the cost of each order in Exercises 15 and 16.

$8.95 $3.99 $10.89 $5.95

15. 4 pairs of flip flops _____

16. 3 sunglasses and 2 pairs of flip flops

_____

Brain Builders

17. **Processes &Practices** **2** **Use Number Sense** About 57 people rent skates at the skating rink each day. It costs $4.25 to rent skates. About how much does the rink make in skate rentals in one week?

_____

18. **Processes &Practices** **4** **Model Math** Write a real-world multiplication problem involving whole numbers and decimals in which you would use estimation to solve. Then estimate to solve the problem. Show that your answer is reasonable.

_____

_____

_____

19. **Building on the Essential Question** Give two examples of why it is important to use estimation when solving problems.

_____

_____

Name ....

MY Homework

Homework Helper

Need help? connectED.mcgraw-hill.com

Estimate the product of 44.5 and 11.

1 Use compatible numbers.

44.5×11

45   10

2 Multiply.

$45 \times 10 = 450$

So, 44.5×11 is about 450.

Practice

Estimate each product.

1. $\$27.64 \times 3$

2. 11.9×21

3. 87.2×41

4. 7.7×4

5. $\$63.44 \times 6$

6. 51.7×9

7. 33.3×33

8. $\$17.55 \times 9$

9. 17.6×51

Problem Solving

10. Processes &Practices

Make Sense of Problems Andrea earns $32.25 a day. After 9 days, about how much will she have earned?

_____

11. Janie skates at the local skating rink for about 2.4 hours every day. If the skating rink is open 54 days a year, about how many hours does Janie skate each year?

_____

12. Each Sunday during his nine-week summer vacation, Ray buys a newspaper. The Sunday paper costs $1.85. About how much does Ray spend on the Sunday newspaper during his vacation?

_____

Brain Builders

13. A group of 5 people goes to the theater. If each ticket costs $6.50 and each person gets a bag of popcorn for $5.75, is $50 a reasonable estimate for the total cost for the group? Explain.

_____

14. An amusement park charges $35.50 for admission. On Saturday, 3,247 people visited the park. On Sunday, 3,542 people visited the park. About how much money did the park earn from admission on Saturday and Sunday?

_____

15. **Test Practice** In Spanish class, Kevin learns an average of 34.5 new words per month. If he takes Spanish for 9 months, which is the best estimate for how many words he will learn?

Ⓐ 350 words   Ⓒ 360 words

Ⓑ 355 words   Ⓓ 365 words

Name _____

Lesson 2
Hands On
Use Models to Multiply

ESSENTIAL QUESTION
How is multiplying and dividing decimals similar to multiplying and dividing whole numbers?

Draw It Tools

Find 0.4 × 2 using decimal models.

1 Shade 4 rows of each model to represent 0.4.

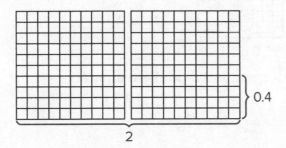

0.4

2

2 Combine both shaded parts onto one model.

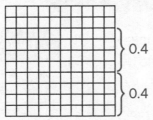

0.4

0.4

The amount shaded is _____. *Eight tenths* of the model is shaded.

So, 0.4 × 2 = _____.

Check Use repeated addition.

$$\begin{array}{r} 0.4 \\ +\ 0.4 \\ \hline 0.8 \end{array}$$

Try It

Find 0.6 × 3 using decimal models.

1 Shade 6 rows of each decimal model to represent 0.6.

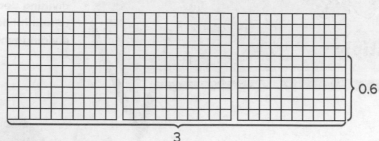

3

0.6

2 Combine all three shaded parts.

Since the combined shaded parts will not fit into one model, use two models.

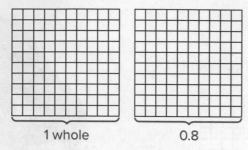

1 whole   0.8

One whole model and *eight-tenths* of a second model are shaded.

The total amount shaded is _____.

So, 0.6 × 3 = _____.

Talk About It

1. The table shows some factors and their products. Study the table. Write a rule you can use to find the product of a whole number and a decimal without using models.

Decimals	Whole Numbers
0.4 × 2 = 0.8	4 × 2 = 8
0.6 × 3 = 1.8	6 × 3 = 18
0.7 × 5 = 3.5	7 × 5 = 35

Processes &Practices **3** **Justify Conclusions** Use your rule from Exercise 1 to find 0.4 × 4 without using models. Explain the process you used.

Practice It

Shade the models to find each product.

3. $0.3 \times 2 =$ _____

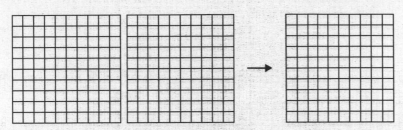

4. $2 \times 0.7 =$ _____

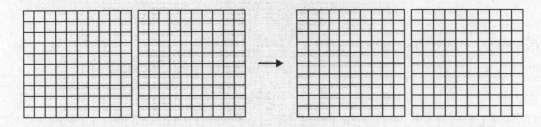

5. $0.3 \times 3 =$ _____

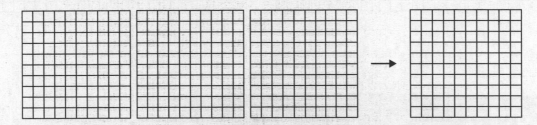

6. $3 \times 0.1 =$ _____

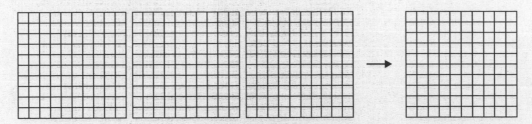

Apply It

Shade decimal models to solve each problem.

Processes &Practices **5** **Use Math Tools** Eva has change to buy bottled
water after gymnastics class for herself and her friend. Each bottle of
water is $0.70. How much change does Eva have?
Use repeated addition.

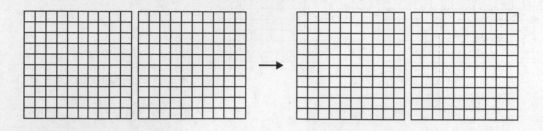

8. Caleb is growing a plant in science class. The plant grows
0.4 centimeters each week. How much will the plant grow in 3 weeks?

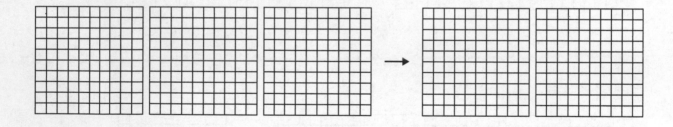

Write About It

9. How does using a model help me multiply decimals by
whole numbers?

MY Homework

Homework Helper

 eHelp

Need help? connectED.mcgraw-hill.com

Find 0.2 × 3 using decimal models.

1 The two rows of each model that are shaded represent 0.2.

2 The shaded parts are combined onto one model.

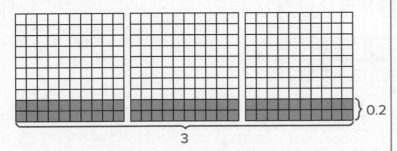

} 0.2

3

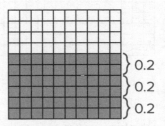

} 0.2
} 0.2
} 0.2

Six-tenths of the model is shaded.

So, 0.2 × 3 = 0.6.

Practice

1. Shade the models to find 3 × 0.7.

3 × 0.7 = _____

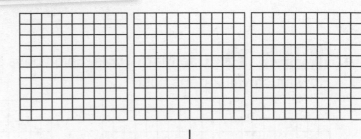

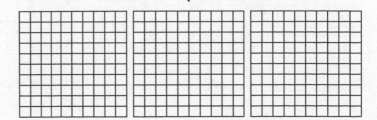

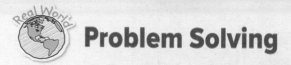

Problem Solving

Shade decimal models to solve each problem.

2. Tyrell lives in a town that is 0.8 kilometers above sea level. He works at a ski resort that is 3 times higher above sea level. How many kilometers above sea level is the ski resort?

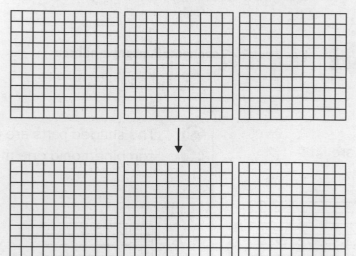

3. Poncio walks 0.5 mile to school each day. Multiply the distance by 2 to find the total distance he walks each day.

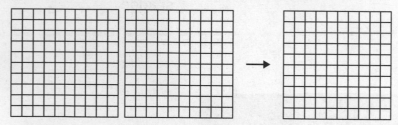

4. **Processes &Practices** **5** **Use Math Tools** Sarah buys a pen that costs $0.75. How much will 2 pens cost?

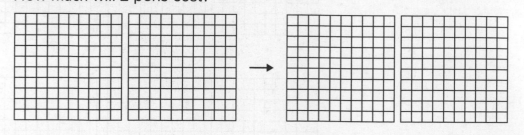

Name

Lesson 3
Multiply Decimals by Whole Numbers

ESSENTIAL QUESTION
How is multiplying and dividing decimals similar to multiplying and dividing whole numbers?

 Math in My World Watch Tutor

Example 1

Find the area of a rectangular board that is **4 feet** by **3.62 feet**.

One Way Use repeated addition.

4 × 3.62 means _____ + _____ + _____ + _____ .

$$
\begin{array}{r}
\overset{2}{3}.6\ 2 \\
3.6\ 2 \\
3.6\ 2 \\
+\ 3.6\ 2 \\
\hline
\square\ \square.\square\ \square
\end{array}
$$

> **Helpful Hint**
> The area of a rectangle is found by multiplying the length by the width.

Another Way Multiply as with whole numbers. Then count decimal places.

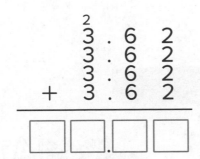

$$
\begin{array}{r}
\overset{2}{3}.62 \\
\times\ \ 4 \\
\hline
14.48
\end{array}
$$

There are _____ places to the right of the decimal point.

Count _____ decimal places from right to left in the product.

Check The answer is the same whether you use repeated addition or multiplication. So, the answer is reasonable.

Example 2

Find 3 × 0.96.

Estimate 3 × 1 = _____

 Multiply as with whole numbers.

> Write the number with the most digits on top. →

$$
\begin{array}{r}
0\ .\ 9\ 6 \\
\times\qquad 3 \\
\hline
\square\ .\ \square\ \square
\end{array}
$$

 Count the decimal places. There are 2 places to the right of the decimal point in 0.96. So, count 2 places from right to left in the product.

So, 3 × 0.96 = _____.

Check The product, _____, is close to the estimate, _____.

The answer is reasonable.

Guided Practice

Multiply. Check for reasonableness.

1.
$$
\begin{array}{r}
\square \\
0\ .\ 5 \\
\times\qquad 6 \\
\hline
\square\ .\ \square
\end{array}
$$

2.
$$
\begin{array}{r}
\square \\
2\ .\ 6 \\
\times\qquad 4 \\
\hline
\square\ \square\ .\ \square
\end{array}
$$

Is the product of 2.8 and 2 greater than 6 or less than 6? How do you know?

392 **Chapter 6** Multiply and Divide Decimals

Copyright © McGraw-Hill Education

Independent Practice

Multiply. Check for reasonableness.

3. 2.49
 × 3

4. 1.59
 × 7

5. 3.4
 × 7

6. $2 \times 1.3 =$ _____

7. $3 \times 0.5 =$ _____

8. $1.8 \times 9 =$ _____

9. 0.48
 × 3

10. 2.4
 × 8

11. 0.02
 × 4

12. $0.66 \times 5 =$ _____

13. $67 \times 4.3 =$ _____

14. $52 \times 2.1 =$ _____

Algebra Find each unknown.

15. $0.8 \times 9 = h$

16. $6 \times 0.05 = x$

17. $1.48 \times 7 = m$

$h =$ _____

$x =$ _____

$m =$ _____

Problem Solving

18. Mr. Thomas's car can travel 17.6 miles per gallon of gas. If his gas tank holds 11 gallons, how far can he travel on a full tank of gas?

_____

19. Tom and Sonia both mowed lawns last week. Tom mowed five lawns and charged $10.75 per lawn. Sonia mowed four lawns but charged $2.25 more per lawn than Tom. Who made the most money?

_____

_____

Brain Builders

20. **Processes &Practices** **1** **Plan Your Solution** Write a multiplication problem in which the product has two decimal places. Alter your problem so that the product has one decimal place.

_____

21. **Processes &Practices** **2** **Use Number Sense** Place the decimal point in one of the factors below to make it correct. Explain your reasoning.

$$498 \times 832 = 4{,}143.36$$

_____

_____

22. **Building on the Essential Question** How is multiplying a decimal by a whole number similar to and different from multiplying two whole numbers? Provide an example to illustrate your reasoning.

_____

_____

Name _____

MY Homework

Lesson 3

Multiply Decimals by Whole Numbers

Homework Helper

Need help? connectED.mcgraw-hill.com

Find 3 × 0.85.

One Way Use repeated addition.

3 × 0.85 means 0.85 + 0.85 + 0.85.

$$
\begin{array}{r}
\overset{2\ 1}{0.85} \\
0.85 \\
+\ 0.85 \\
\hline
2.55
\end{array}
$$

Another Way Multiply as with whole numbers. Then count decimal places.

$$
\begin{array}{r}
\overset{2\ 1}{0.85} \\
\times\ \ \ 3 \\
\hline
2.55
\end{array}
$$

← There are two places to the right of the decimal point.

Count two decimal places from right to left in the product.

Practice

Multiply. Check for reasonableness.

1. $\begin{array}{r} 1.7 \\ \times\ 4 \\ \hline \end{array}$

2. $\begin{array}{r} 0.62 \\ \times\ \ 2 \\ \hline \end{array}$

3. $\begin{array}{r} 0.5 \\ \times\ 9 \\ \hline \end{array}$

4. 3.6 × 8 = _____

5. 5.1 × 7 = _____

6. 4 × 2.3 = _____

Algebra Find each unknown.

7. 2 × 0.33 = p

p = _____

8. 5 × 2.4 = n

n = _____

9. 7 × 8.1 = s

s = _____

10. Justin bought 9 packs of trading cards. Each pack of cards costs $3.27. What was the total cost for all 9 packs?

_____

11. Richard paints a picture on a rectangular canvas that is 3 feet by 2.64 feet. What is the area of Richard's painting?

_____

Brain Builders

12. **Processes &Practices** **6** **Explain to a Friend** Taylor wants to buy a bicycle that costs $48.75 and a helmet that costs $18.30. She works for her mom after school to earn the money. If Taylor is paid $5.50 per hour, will she be able to buy the bicycle and helmet after working 9 hours? Explain to a friend.

_____

_____

13. Janita practices on the track field 1.5 hours each day. She needs to complete 12 hours of practice before the next meet. She has already practiced for 6 days. How many more hours does Janita need to practice?

_____

14. **Test Practice** Kent buys a pen that costs $1.21. How much will 4 pens cost?

Ⓐ $3.63   Ⓒ $4.21

Ⓑ $3.84   Ⓓ $4.84

Name _____

Lesson 4
Hands On
Use Models to Multiply Decimals

ESSENTIAL QUESTION
How is multiplying and dividing decimals similar to multiplying and dividing whole numbers?

Draw It Tools

Find 0.3 × 0.7 using a decimal model.

 1 Use one 10-by-10 decimal model.

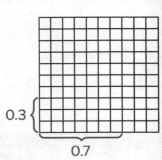

2 Shade a rectangle that is 0.3 unit wide and 0.7 unit long.

3 There are _____ hundredths in the shaded region. The shaded

region represents the area of a rectangle with a length of 0.7 unit and a

width of 0.3 unit. So, 0.3 × 0.7 = _____ .

Talk About It

1. How are the equations 3 × 7 = 21 and 0.3 × 0.7 = 0.21 similar? How are they different?

_____

_____

2. **Processes &Practices** **5** **Use Math Tools** How is using a model to multiply 0.3 by 0.7 similar to finding the area of a rectangle?

_____

_____

Try It

Find 0.4 × 2.4 using decimal models.

1 Use three 10-by-10 decimal models side by side.

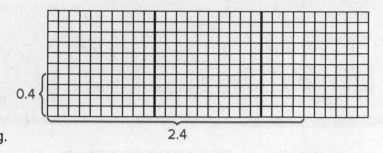

2 Shade a rectangle that is 0.4

0.4 {

unit wide and _____ units long.

2.4

3 Shade the same amount of squares on a 10-by-10 grid.

4 There are _____ hundredths in the shaded region.

So, 0.4 × 2.4 = _____ .

Talk About It

3. Without using models, explain why 2 tenths times 3 tenths is equal to 6 *hundredths*. Use place value in your explanation.

_____

4. The table shows some factors and their products. Study the table. Write a rule you can use to find the product of two decimals, both to the tenths place, without using models.

Decimals	Whole Numbers
0.3 × 0.7 = 0.21	3 × 7 = 21
0.4 × 2.4 = 0.96	4 × 24 = 96
0.5 × 1.1 = 0.55	5 × 11 = 55

_____

Processes
5. **&Practices** **▶2** **Stop and Reflect** Without using models, find 0.4 × 0.8. Explain how you found the product.

_____

_____

Practice It

Shade the decimal models to find each product.

6. $0.4 \times 0.8 =$ _____

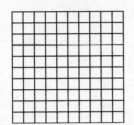

7. $0.5 \times 0.6 =$ _____

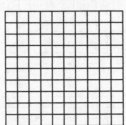

8. $0.2 \times 0.9 =$ _____

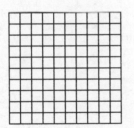

9. $0.4 \times 0.6 =$ _____

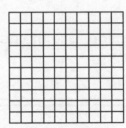

10. $0.3 \times 1.8 =$ _____

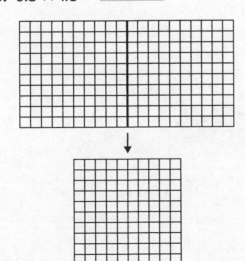

11. $0.2 \times 1.4 =$ _____

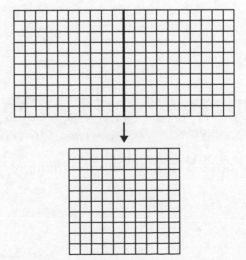

12. Diana's house has a rectangular window on the door with a height of 1.5 feet and a width of 0.8 feet. What is the area of the window? Shade the decimal models to solve the problem.

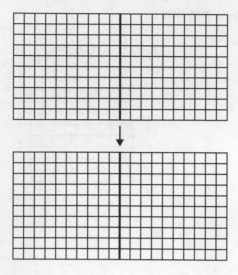

**Processes
13. & Practices** **2** **Reason** Use the decimal model below to explain why 0.5×0.11 is 0.055.

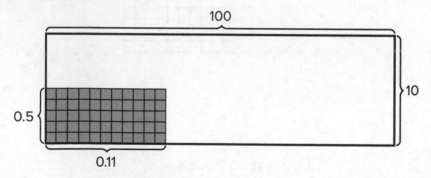

Write About It

14. How does a model help me multiply decimals?

Name _____

 **MY Homework**

Homework Helper

Need help? ➚ **connectED.mcgraw-hill.com**

Find 0.4 × 0.6 using a decimal model.

1 Use one 10-by-10 decimal model.

2 The shaded rectangle represents 0.4 unit wide and 0.6 unit long.

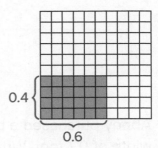

0.4 { ⟍
0.6

3 There are twenty-four hundredths in the shaded region.

So, 0.4 × 0.6 = 0.24.

Practice

Shade the decimal models to find each product.

1. 0.7 × 0.5 = _____

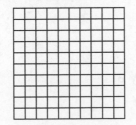

2. 0.8 × 0.2 = _____

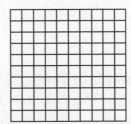

3. 0.6 × 2.2 = _____

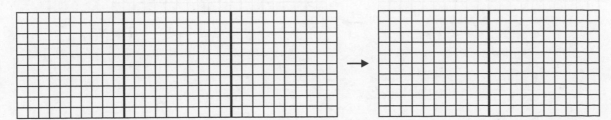

Problem Solving

Shade the decimal models to solve each problem.

4. Lanny bought 1.4 pounds of rice that cost $0.40 per pound. How much does Lanny pay for the rice?

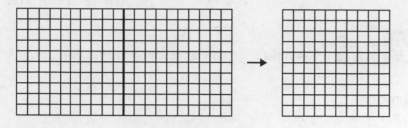

_____

5. Abbey purchased a poster that had a height of 2 feet and a width of 0.7 foot. What is the area of Abbey's poster?

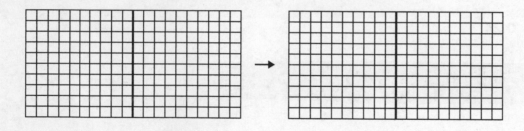

_____

6. **Processes & Practices** **5** **Use Math Tools** Diana spends 0.8 hour per day at the pool. If she goes to the pool for 2 days one week, how many total hours does Diana spend at the pool that week?

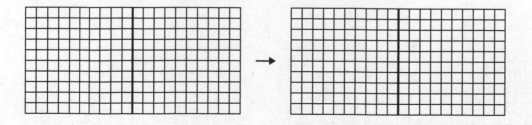

_____

Lesson 5
Multiply Decimals

ESSENTIAL QUESTION
How is multiplying and dividing decimals similar to multiplying and dividing whole numbers?

 Math in My World Watch ▶ Tutor 💬

Example 1

Nina is buying peanuts in bulk. The peanuts cost **$0.75** for each pound. She purchased 2.1 pounds of peanuts. How much will Nina pay? Round to the nearest cent.

Find $0.75 × 2.1.

Estimate $0.75 × 2.1 ⟶ _____ × _____ or _____

1 Multiply as with whole numbers.

2 Count the decimal places.

3 Since 2 + 1 = 3, count 3 decimal places from the right to place the decimal point.

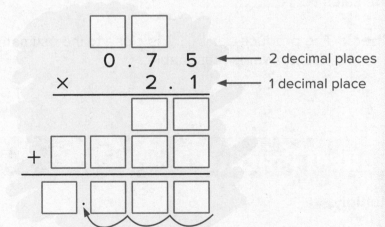

0 . 7 5 ⟵ 2 decimal places
× 2 . 1 ⟵ 1 decimal place

4 Round to the nearest cent.

$1.575 ⟶ _____

So, Nina will need to pay _____ .

Check $2 ≈ _____

Example 2

Tutor

Demarcus is buying a new guitar pick that costs $0.50. The sales tax is found by multiplying the cost of the guitar pick by 0.06. What is the cost of the sales tax for the pick?

Find 0.50×0.06.

Estimate $0.50 \times 0.06 \longrightarrow \$1 \times 0 = \$0$

 Multiply as with whole numbers.

 Count the decimal places.

 Since $2 + 2 = 4$, count 4 decimal places from the right. Annex a zero to place the decimal point.

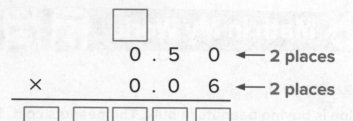

$0 . 5 \quad 0 \longleftarrow$ **2 places**

$\times \qquad 0 . 0 \quad 6 \longleftarrow$ **2 places**

So, $0.50 \times 0.06 =$ _____ .

The sales tax is _____ .

Check The product, _____ , is close to the estimate, _____ .
The answer is reasonable.

Guided Practice

Multiply.

1. $0.6 \times 0.8 =$ _____

2. $1.7 \times 2.4 =$ _____

3. $0.9 \times 3.8 =$ _____

Describe a multiplication problem in which the product is between 0.005 and 1.

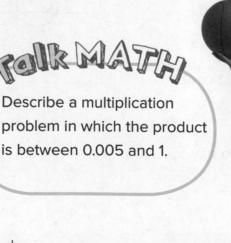

Independent Practice

Multiply. Check for reasonableness.

4. 0.96
   × 7.1

5. 3.65
   × 2.6

6. 0.07
   × 5.2

7. 2.78
   × 0.8

8. 0.35
   × 0.15

9. 3.24
   × 6.4

10. 0.9 × 0.3 = _____

11. 1.6 × 3.2 = _____

12. 0.5 × 6.7 = _____

Algebra Find each unknown.

13. 0.81 × 7.3 = b

   b = _____

14. 5.6 × 3.9 = p

   p = _____

15. 1.2 × 0.05 = g

   g = _____

16. A nutrition label says that one serving out of a bag of chips has 3.7 grams of fat. How many grams of fat are there in 2.5 servings?

_____

17. Apples cost $0.98 per pound. How much would it cost to purchase 5.5 pounds of apples?

_____

18. The average honeybee can travel 10.5 feet per second. How many feet can the honeybee travel in 3.7 seconds?

_____

Brain Builders

19. **Processes &Practices** **6** **Be Precise** Jocelyn's swimming pool is shown. The community center swimming pool is 200 square feet larger than Jocelyn's pool. Find the area of the community center swimming pool. Show your work.

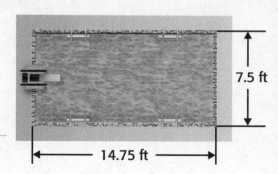

7.5 ft

14.75 ft

_____

_____

20. **Processes &Practices** **4** **Model Math** Write two different multiplication problems using two decimals that have a product greater than 0.1 but less than 0.2.

_____

21. **Building on the Essential Question** How is multiplying two decimals different from multiplying a whole number and a decimal?

_____

_____

Name

MY Homework

Homework Helper

Need help? *connectED.mcgraw-hill.com*

Find 0.32 × 0.3.

Estimate 0.32 × 0.3 ⟶ 0 × 0 = 0

1 Multiply as with whole numbers.

2 Count the decimal places.

3 Since 2 + 1 = 3, count 3 decimal places from the right. Annex a zero to place the decimal point.

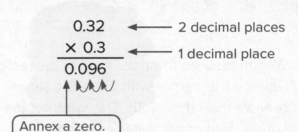

So, 0.32 × 0.3 = 0.096.

Check The product, 0.096, is close to the estimate, 0. The answer is reasonable.

Practice

Multiply. Check for reasonableness.

1. 0.28
    × 0.03

2. 0.11
    × 0.91

3. 5.14
    × 0.4

4. 0.98 × 0.23 =

5. 0.52 × 0.48 =

6. 6.34 × 0.7 =

Algebra Find each unknown.

7. $0.47 \times 0.18 = t$

t = _____

8. $4.15 \times 6.3 = x$

x = _____

9. $0.9 \times 1.02 = w$

w = _____

Problem Solving

10. Tamara recycles 29.5 pounds of aluminum cans at a recycling center that pays $0.26 per pound. How much does Tamara receive for the cans?

_____

Brain Builders

11. Wes measures the floor area of his rectangular living room to replace the carpet with tiles. The length of the room is 2.6 feet greater than the width. The width of the room is 9.8 feet. How many square feet of tile will Wes need?

_____

12. **Processes &Practices** **2** **Use Number Sense** Lenora buys 3.7 pounds of potatoes that cost $0.29 per pound and 4.8 pounds of onions that cost $1.23 per pound. How much does Lenora pay in all? Round to the nearest cent.

_____

13. **Test Practice** Mikayla is buying deli meat and cheese for sandwiches for a family reunion. How much will she spend on 4.5 pounds of ham and 3.5 pounds of cheese?

ham
$4.59
per pound

cheese
$3.79
per pound

ⓐ $7.39

ⓒ $20.66

ⓑ $13.27

ⓓ $33.92

Check My Progress

Vocabulary Check

Draw lines to match each word(s) with its correct description or meaning.

1. compatible numbers

2. estimate

3. product

- an approximate value

- the result of a multiplication problem

- numbers that are easy to compute mentally

Concept Check

Estimate each product.

4. $1.80 × 8

5. $2.83 × 7

_____

Multiply. Check for reasonableness.

6. 1.9
    × 8

7. 3.4
    × 7

8. 2.3 × 2 = _____

9. 0.24 × 5 = _____

Problem Solving

10. Jorge bought 8 pounds of ground beef for $3.29 a pound. About how much did he pay altogether?

_____

11. The Marino family bought 4 tickets to the circus. What was the total cost of the tickets?

_____

Brain Builders

12. Wendy paints 4 rectangular walls that are 10.5 feet tall and 9.2 feet wide. What is the total area of the walls that she paints? Explain.

_____

_____

13. One pound of tomatoes costs $1.59. One pound of bananas costs $1.87. How much more do 5 pounds of tomatoes cost than 2 pounds of bananas? Show your work.

_____

_____

14. Test Practice Mr. Garner's gas tank holds 17.5 gallons of gasoline. How much will it cost him to fill up his gas tank if gasoline costs $2.48 per gallon assuming there is no gas in the tank to start with?

Ⓐ $34   Ⓒ $43.40

Ⓑ $45.20   Ⓓ $42.16

Lesson 6
Multiply Decimals by Powers of Ten

 Math in My World [Watch ▶] [Tutor 💬]

Example 1

Josh will need to make 10 payments of $32.25 to purchase a new skimboard. What is the total cost of the skimboard?

Find 32.25×10.

$$
\begin{array}{r}
32.25 \\
\times\ 10 \\
\hline
0000 \\
+\ 32250 \\
\hline
322.50
\end{array}
$$

> The non-zero digits are the same. The decimal point in the product moved one place to the right.

Multiplying a decimal by 10 moves the decimal point one place to the right.

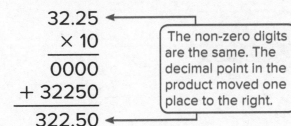

32.25 ⟶ 322.50

So, the skimboard cost _____.

Numbers like 10, 100, and 1,000 are powers of ten. They can be written with exponents with a base of ten.

There is 1 zero in 10. The exponent on 10^1 is 1.

There are 2 zeros in 100. The exponent on 10^2 is 2.

There are 3 zeros in 1,000. The exponent on 10^3 is 3.

Power of Ten	Written with Exponent
10	10^1
100	10^2
1,000	10^3

To multiply a decimal by a power of ten, move the decimal point to the right the same number of zeros in the power of ten. This is also the same number as the exponent on 10.

Example 2

Find 24.7 × 10².

One Way Multiply.

Since $10^2 = 100$, find 24.7 × 100.

$$
\begin{array}{r}
24.7 \\
\times 100 \\
\hline
000 \\
0000 \\
+ \ 24700 \\
\hline
2{,}470.0
\end{array}
$$

The non-zero digits are the same. The decimal point in the product moved two places to the right.

So, 24.7 × 10² = _____ .

Another Way Move the decimal point.

Multiplying a decimal by 100 moves the decimal point two places to the right.

Annex a zero.

24.70 ⟶ 2,470

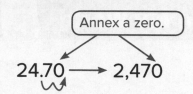

Guided Practice

Multiply.

1. 0.54 × 10 = _____

2. 8.32 × 100 = _____

3. 7.46 × 1,000 = _____

Talk MATH

Explain how you can mentally find the cost of 10 text messages that each cost $0.25.

Independent Practice

Multiply.

4. $1.63 \times 10 =$ _____

5. $0.853 \times 10^3 =$ _____

6. $0.397 \times 10^1 =$ _____

7. $1.76 \times 100 =$ _____

8. $0.78 \times 10^2 =$ _____

9. $76.5 \times 10^3 =$ _____

10. $0.81 \times 10 =$ _____

11. $1.23 \times 10^2 =$ _____

12. $0.48 \times 100 =$ _____

Algebra Find each unknown.

13. $0.93 \times 10 = a$

$a =$ _____

14. $22.94 \times 10^2 = n$

$n =$ _____

15. $0.05 \times 1,000 = w$

$w =$ _____

16. A music store has 10 flutes. Each flute costs $325.50. What is the cost of all 10 flutes?

_____

17. **Processes &Practices**  ➊ **Make a Plan** One carton of milk costs $0.99. What is the total cost of 10^2 cartons of milk?

_____

18. Titus's and Cam's hourly charges for yard work are shown. Suppose Titus and Cam each worked 10 hours. How much money did they earn together?

Titus	Cam
$8.25	$5.58

Brain Builders

19. Jenny's hamster is 8.4 centimeters long. Her dog is 10 times as long as her hamster, and her cat is 4 times as long as her hamster. How much longer is her dog than her cat?

_____

20. **Processes &Practices** ➋ **Use Number Sense** Find the missing exponent.
$0.346 \times 10^{?} = 34.6$.

_____

21. **Building on the Essential Question** How does the exponent of each power of ten correspond with placing a decimal?

_____

_____

Name

MY Homework

Lesson 6
Multiply Decimals by Powers of Ten

Homework Helper

Need help? connectED.mcgraw-hill.com

Find 0.35 × 100.

One Way Multiply.

$$
\begin{array}{r}
0.35 \\
\times\ 100 \\
\hline
000 \\
0000 \\
+\ 3500 \\
\hline
35.00
\end{array}
$$

The non-zero digits are the same. The decimal point in the product moved two places to the right.

Another Way Move the decimal point.

Multiplying a decimal by 100 moves the decimal point two places to the right.

0.35 ⟶ 35

So, 0.35 × 100 = 35.

Practice

Multiply.

1. 0.13 × 10 = _____

2. 1.4 × 1,000 = _____

3. 4.81 × 10^3 = _____

4. 0.72 × 10^2 = _____

5. 0.179 × 1,000 = _____

6. 67.2 × 10^1 = _____

Copyright © McGraw-Hill Education

![Real World] **Problem Solving**

For Exercises 7 and 8, use the table which shows a school store's prices.

Item	Price
Notebook	$1.25
Pencil	$0.50
Binder	$2.15
Pen	$0.80

7. How much would it cost to buy 10 pens from the school store?

_____

8. How much will it cost for 10 pencils and one notebook?

_____

Brain Builders

9. The school budgeted $450 to purchase trophies to give to the honor roll students. If one trophy costs $4.32, how much money will be left over after purchasing 100 trophies?

_____

10. **Processes &Practices** **3** **Find the Error** Rico is finding 7.5 × 100. Find and correct his mistake. Explain how Rico could have known that his answer was incorrect.

$$7.5 \times 100$$
$$07.5 = 0.075$$

_____

_____

_____

11. **Test Practice** Elyse is cutting strips of paper for a scrapbook page. Each strip is 1.5 centimeters wide. How wide will 10 strips of paper be?

Ⓐ 1.5 centimeters   Ⓒ 150 centimeters

Ⓑ 15 centimeters   Ⓓ 1,500 centimeters

Lesson 7
Problem-Solving Investigation
STRATEGY: Look For a Pattern

ESSENTIAL QUESTION
How is multiplying and dividing decimals similar to multiplying and dividing whole numbers?

Learn the Strategy

 Watch   Tutor

The table shows the amount Hank pays for his clothes and the amount of his discount. How much will he pay, and how much will the discount be for a shirt that costs $34.70?

Item	Cost	Discount Amount	Final Cost
Boots	$68.80	$6.88	$61.92
Jacket	$62.30	$6.23	$56.07
Pants	$48.50	$4.85	$43.65
Shirt	$34.70	■	■

1 Understand

What facts do you know?

- Hank will receive a _____ on his _____ .

What do you need to find?

- I need to find the amount Hank will pay

  for the _____ .

2 Plan

Look for a pattern. Extend the pattern to find the final cost.

3 Solve

Multiply the cost of the item by _____ to find the discount.

Subtract the discount from the _____ to find the final cost.

$34.70 × _____ = _____   The discount is _____ .

$34.70 − _____ = _____   Hank will pay _____ .

4 Check

Does your answer make sense? Explain.

_____

Adrienne did 13 sit-ups the first day, 20 sit-ups the second day, and 27 sit-ups the third day. If the pattern continues, how many sit-ups will she do on the sixth day?

 Understand

What facts do you know?

_____

_____

What do you need to find?

_____

 **Plan**

_____

_____

❸ Solve

❹ Check

Is my answer reasonable?

_____

Apply the Strategy

Solve each problem by looking for a pattern.

1. Every year, Victoria receives $30 for her birthday, plus $2 for each year of her age. Lacey receives $20 for her birthday and $4 for each year of her age. In 2013, Victoria is 10, and Lacey is 6. In what year will they both receive the same amount of money?

_____

2. Trent lifts weights 7 days a week. He spends 18 minutes lifting weights on Monday, 29 minutes on Tuesday, 40 minutes on Wednesday, and 51 minutes on Thursday. If this pattern continues, how many minutes will Trent lift weights on Saturday?

Brain Builders

3. Find the missing numbers in the table. Then describe the pattern.

_____

_____

_____

Input	Output
5	9
10	19
15	■
25	49
■	79

4. **Processes &Practices** ▶**8** **Look for a Pattern** Describe the pattern below. Then find the missing numbers.

0.03, 0.3, 3, 30, _____, _____, _____

_____

5. Describe the pattern below. Then find the missing numbers.

784.5, 78.45, 7.845, _____, _____ .

_____

Review Strategies

Use any strategy to solve each problem.
- Make a table.
- Choose an operation.
- Act it out.
- Draw a picture.

Processes &Practices ➊ **Plan Your Solution** Samuel will arrive at the airport on the first plane after 10 A.M. Airplanes arrive every 50 minutes beginning at 6 A.M. When will Samuel's plane arrive?

7. On Serena's tenth birthday, her mom was 3 times Serena's age. How old will Serena and her mom be when her mom's age times 0.5 will equal Serena's age?

8. Mindy read 8 pages of her book the first day, 15 pages the second day, and 22 pages the third day. If the pattern continues, how many pages of her book will she read on the sixth day?

9. A 1-mile hiking path has signs placed every 240 feet. There are signs placed at the beginning and end of the mile. How many signs are there? (*Hint:* 1 mile = 5,280 feet)

10. Jada lives in a city that has an area of 344.6 square miles. Her friend lives in a town that is one-tenth, or 0.1, that size. What is the area of her friend's town?

11. Dennis has 9.5 weeks to prepare for a bike tour. If he rides 8.2 miles each week, how many miles will he ride before the bike tour?

Name ....

Homework Helper

Need help? *connectED.mcgraw-hill.com*

Deidra wants to buy new summer clothes. She
needs to find the total cost in order to
determine how much she can afford. The
table shows the prices of some items. Based
on the pattern, what is the sales tax and total
cost for jeans that are $18.00? Round each
amount to the nearest cent.

Item	Price	Sales Tax	Total Cost
T-shirt	$8.00	$0.56	$8.56
Shorts	$10.00	$0.70	$10.70
Sandals	$15.00	$1.05	$16.05
Jeans	$18.00	■	■

1 Understand

What facts do you know?

The price, sales tax, and total cost of the T-shirt, shorts, and sandals.

The price of the jeans.

What do you need to find?

The sales tax and total cost of the jeans.

2 Plan

Look for a pattern to solve the problem.

3 Solve

Multiply the cost of the item by 0.07 to find the sales tax, rounded
to the nearest cent. Add the sales tax to the price of the item.

$18.00 × 0.07 = $1.26   The sales tax is $1.26.

$18.00 + $1.26 = $19.26   The total cost is $19.26.

4 Check

$19.26 − $1.26 = $18.00, so the answer is correct.

Problem Solving

Solve each problem by looking for a pattern.

1. Harold has 4 classes each school morning. Each class is 1 hour long, and there are 10 minutes between classes. The first class starts at 8:00 A.M. What time does the fourth class end?

   _____

2. Mr. Moore read 25 pages of a book on Monday. He read 32 pages on Tuesday, 39 pages on Wednesday, and 46 pages on Thursday. If this pattern continues, how many pages will Mr. Moore read on Sunday?

   _____

Brain Builders

3. **Processes &Practices** **8** **Look for a Pattern** Which number does not belong to the pattern below? Explain.
   5,470, 547, 54.7, 0.547

   _____

4. Draw the next two figures in the pattern. Describe the pattern.

   _____

   _____

5. Marissa can buy 3 oranges for $1.35, 4 oranges for $1.80, or 5 oranges for $2.25. Marissa can by 10 oranges for a price that fits this pattern or a bag of 10 oranges for $4.25. Which way costs less for 10 oranges? Explain.

   _____

   _____

   _____

Lesson 8
Multiplication Properties

ESSENTIAL QUESTION
How is multiplying and dividing decimals similar to multiplying and dividing whole numbers?

 Math in My World Tutor

Example 1

A coach had 16 players in each of 2 groups. Each player scored 5 goals. Find the total number of goals scored.

Find (16 × 2) × 5.

Since you can easily multiply 2 and 5, change the way the numbers are grouped.

The **Associative Property of Multiplication** states that the way in which factors are grouped does not change the product.

It is easier to find 2 × 5 than 16 × 2. You can group the numbers differently to find 2 × 5 first.

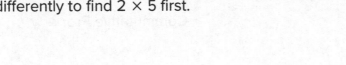

(16 × 2) × 5 = 16 × (2 × 5)   Associative Property

= 16 × _____   Multiply. Parentheses tell you which factors to multiply first.

= _____   Multiply.

So, the total number of goals scored is _____.

Example 2

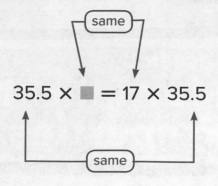

Find the unknown in the equation
35.5 × ■ = 17 × 35.5.

The **Commutative Property of Multiplication** shows
that the order in which factors are multiplied does not
change the product.

So, the unknown is _____ .

Example 3

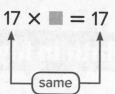

Find the unknown in the equation 17 × ■ = 17.

The **Identity Property of Multiplication** states that the
product of any factor and 1 equals the factor.

So, the unknown is _____ .

Guided Practice

Draw lines to match the multiplication property used in each equation.

1. 6.2 × 100 = 100 × 6.2

2. (8 × 2) × 3 = 8 × (2 × 3)

3. 78.56 × 1 = 78.56

• Identity Property

• Commutative Property

• Associative Property

Talk MATH

Explain how you could
use mental math and
multiplication properties
to find (5.5 × 50) × 2.

Independent Practice

Use properties of multiplication to find each product mentally. Show your steps and identify the properties that you used.

4. $(5.1 \times 2) \times 50 =$ _____

5. $4 \times (2.5 \times 6) =$ _____

6. $(9.8 \times 500) \times 2 =$ _____

7. $(1.4 \times 50) \times 20 =$ _____

Processes &Practices **2** **Use Algebra** **Find the unknown in each equation. Circle which property you used.**

8. $19.5 \times \blacksquare = 19.5$

$\blacksquare =$ _____

Commutative Property

Associative Property

Identity Property

9. $34 \times 65 = 65 \times \blacksquare$

$\blacksquare =$ _____

Commutative Property

Associative Property

Identity Property

10. $2.1 \times \blacksquare = 4.3 \times 2.1$

$\blacksquare =$ _____

Commutative Property

Associative Property

Identity Property

11. $(17 \times 2) \times 5 = 17 \times (\blacksquare \times 5)$

$\blacksquare =$ _____

Commutative Property

Associative Property

Identity Property

Problem Solving

12. Elijah and 2 of his friends are each paid $20 per afternoon for stuffing envelopes. If they work 5 afternoons, what is the total amount of their earnings?

_____

13. Replace the ■ in (8.7 × ■) × 5 with a number greater than 10 so that the product is easy to find mentally. Explain.

_____

_____

_____

Brain Builders

14. **Processes &Practices** ▶**3** **Draw a Conclusion** Each juice box contains 6.4 ounces. Each value pack of juice holds 10 juice boxes. If you have fifty value packs, do you have 4,000 ounces of juice? Explain.

_____

15. **Processes &Practices** ▶**2** **Use Number Sense** Write a problem that you could solve mentally using two multiplication properties. Explain how the properties help you solve the problem.

_____

_____

16. **?** **Building on the Essential Question** How do the multiplication properties help me to find products mentally? Provide an example to illustrate your explanation.

_____

_____

Name ....

MY Homework

Homework Helper

Need help? connectED.mcgraw-hill.com

Use properties of multiplication to find (1.7 × 5) × 2.

(1.7 × 5) × 2 = 1.7 × (5 × 2)   Associative Property

= 1.7 ×   10   Multiply.

=   17   Multiply.

So, (1.7 × 5) × 2 = 17.

Practice

Use properties of multiplication to find each product mentally.
Show your steps and identify the properties that you used.

1. (1.6 × 2) × 5 = _____

2. (27 × 2.5) × 4 = _____

Algebra Find each unknown.

3. ■ × 5.5 = 5.5

4. 49 × 201 = ■ × 49

■ = _____

■ = _____

Problem Solving

5. For a party, Shandra and James each bought 5 packages of hot dog buns that each cost $1.50. How much did the hot dog buns cost altogether?

_____

6. Eva needs to find: (6 × 1.5) × 2. Give the product and name the property she could use.

_____

Brain Builders

7. Petra claims that he can use the Associative Property of Multiplication to rewrite (9.5 × 9) + 1 as 9.5 × 10. Is Petra correct? Explain.

_____

_____

8. Processes &Practices **2** **Reason** Without calculating, is the equation (1.8 × 3) × (2.1 × 7) = (3 × 7) × (1.8 × 2.1) true or false? Explain your reasoning.

_____

_____

9. Test Practice Which multiplication property is shown in the equation below?

3.1 × (2 × 9) = (3.1 × 2) × 9

Ⓐ Identity Property

Ⓑ Commutative Property

Ⓒ Zero Property

Ⓓ Associative Property

Name ....

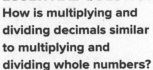

ESSENTIAL QUESTION
How is multiplying and dividing decimals similar to multiplying and dividing whole numbers?

 Math in My World Watch ▶ Tutor 💬

Example 1

Ms. Glover buys 2 kites for a total of $15.18. If each kite costs the same, about how much did each kite cost? Explain why your answer is reasonable.

Estimate the quotient of 15.18 and 2.

 Use compatible numbers.

$$\$15.18 \div 2$$

⬇   ⬇

[ ] ÷ 2

⬇   ⬇

> **Helpful Hint**
> Compatible numbers are numbers that are easy to compute mentally.

 Divide. _____ ÷ 2 = _____

So, $15.18 ÷ 2 is about _____ .

Since 2 × 8 = _____ and _____ ≈ 15.18, the answer is reasonable.

> The ≈ means *approximately equal to.*

Example 2

Three friends went out to dinner. The total cost of their meals was $32.57. If the friends split the bill evenly, about how much will each person pay? Explain why your answer is reasonable.

Estimate $32.57 ÷ 3 by rounding.

 Round $32.57 to the nearest whole dollar because 33 and 3 are compatible numbers.

$32.57 ÷ 3

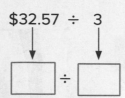

Divide.

_____ ÷ _____ = _____

Each friend will pay about _____.

Since 3 × _____ = _____ and

_____ ≈ 32.57, the answer is reasonable.

Guided Practice

Estimate each quotient.

1. $19.50 ÷ 5

☐ ÷ ☐ = ☐

Talk MATH

Describe another way you could estimate in Example 2. Are both estimates reasonable? Explain.

2. $47.25 ÷ 25

☐ ÷ ☐ = ☐

3. 16.8 ÷ 4

☐ ÷ ☐ = ☐

Independent Practice

Estimate each quotient.

4. 87.3 ÷ 11

5. 44.7 ÷ 5

6. 195.8 ÷ 12

7. $28.20 ÷ 6

8. $7.92 ÷ 6

9. 88.3 ÷ 9

10. 128.9 ÷ 12

11. 576.4 ÷ 62

12. $15.47 ÷ 7

13. 56.3 ÷ 18

Problem Solving

14. It costs $158.75 to purchase 15 tickets to the state fair. About how much does one ticket cost?

15. Jake bought 3 drawing pens for $8.07. Each pen costs the same amount. About how much did he spend for one pen?

16. A canoe rental company offers two trips along the river. Neil and Gabriela choose the longer trip. If they split the cost of the rental, about how much will each person pay?

Canoe Rental		
Trip	Distance (miles)	Cost
A	5.75	$12.98
B	8.5	$16.32

Brain Builders

17. **Processes &Practices** ➊ **Make Sense of Problems** It costs $96.50 for 6 adults and 1 child to see an exhibit on Egyptian mummies. Child tickets are $8.00. About how much does it cost one adult to see the exhibit?

18. **Processes &Practices** ➍ **Model Math** Write a real-world estimation problem involving dividing a decimal by a whole number. Then estimate the quotient. Explain whether the estimation is greater than or less than the actual amount.

19. **Building on the Essential Question** How can I use compatible numbers to estimate the quotient of a decimal and whole number?

Name _____

MY Homework

Homework Helper

Need help? ↗ **connectED.mcgraw-hill.com**

Estimate 78.74 ÷ 42.

Estimate 78.74 ÷ 42 by rounding.

1 Round 78.74 and 42 to the nearest ten.

$$78.74 \div 42$$

$$80 \div 40 \quad \longleftarrow \quad \boxed{\text{80 and 40 are compatible numbers.}}$$

2 Divide. 80 ÷ 40 = 2

So, 78.74 ÷ 42 is about 2.

Practice

Estimate each quotient.

1. 32.17 ÷ 7

2. $175.32 ÷ 3

3. 21.9 ÷ 3

4. 36.3 ÷ 6

5. 17.5 ÷ 5

6. 120.6 ÷ 2

![Real World] **Problem Solving**

Use the table to answer Exercises 7–9.

The information in the table can be used to determine the density of each object. Density describes how tightly the particles in an object are packed together. You can find density by dividing an object's mass by its volume.

Substance	Mass (g)	Volume (cm³)
Aluminum	13.5	5
Gold	56.7	3
Mercury	121.5	9

7. Estimate the density of aluminum.

8. Is the density of gold greater than the density of mercury? Explain.

9. The density of gold is about how many times greater than the density of aluminum?

Brain Builders

10. Harry's mother makes cakes for a local restaurant. She buys flour and sugar in large amounts. She has 157.86 pounds of flour and 82.69 pounds of sugar. If she uses 15 pounds of flour and 8 pounds of sugar each day, for about how many days will they both last?

11. Processes & Practices ▷4 **Model Math** Write a division problem involving the division of a decimal by a whole number with an estimated quotient of 7. Compare the estimate to the actual quotient.

12. Test Practice Kendall measured the rainfall in her area for a year. Her readings totaled 34.56 inches. Which is the best estimate of the average rainfall per month?

Ⓐ about 1 inch per month

Ⓑ about 2 inches per month

Ⓒ about 3 inches per month

Ⓓ about 4 inches per month

Check My Progress

Vocabulary Check

Identify the multiplication property for each equation.

Associative Property   **Commutative Property**   **Identity Property**

1. $1.5 \times 2 = 2 \times 1.5$ _____

2. $5.5 \times (3 \times 10) = (5.5 \times 3) \times 10$ _____

3. $7.8 \times 1 = 7.8$ _____

Concept Check

Find each product.

4. $0.87 \times 10 =$ _____   **5.** $0.742 \times 100 =$ _____

6. $0.22 \times 10^2 =$ _____   **7.** $1.5 \times 10^3 =$ _____

Describe each pattern below. Then find the next two numbers.

8. 0.03, 0.3, 3, 30, ■, ■   **9.** 784.5, 78.45, 7.845, ■, ■

_____   _____

Estimate each quotient.

10. $12.5 \div 5$ **11.** $17.2 \div 8$ **12.** $23.7 \div 6$

13. Neal is saving $4.25 a week for 100 weeks. How much will he have saved altogether?

14. Five friends went to the movies. The total cost was $55.91. If the friends split the cost equally, about how much did each person pay?

Brain Builders

15. Wendy is going to paint two walls that are each 10.5 feet tall and 9.2 feet wide. She only has 2 cans of paint, and each can covers 50 square feet. Exactly how much of the area will remain unpainted?

16. Kevin swims 50 meters in 2.3 minutes on his first day of swimming class. He swims the same distance in 2.1 minutes a week later. The next week, he swims the same distance in 1.9 minutes. If the pattern continues, what will be Kevin's time two weeks later?

17. Test Practice The Weston Laundry washes all the linens for local hotels. In 7 days, they washed 285.38 pounds of towels and 353.47 pounds of sheets. About how many pounds of laundry did they wash each day?

Ⓐ 80 pounds   Ⓒ 120 pounds

Ⓑ 100 pounds   Ⓓ 140 pounds

Name

Hands On

Divide Decimals

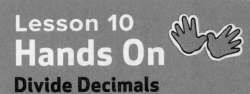

ESSENTIAL QUESTION
How is multiplying and
dividing decimals similar
to multiplying and
dividing whole numbers?

You can use models to divide a decimal by a whole number.

Build It

Find 3.6 ÷ 3 using models.

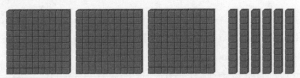

1 The model shows 3.6 using three wholes and six tenths.

2 Divide the blocks into three equal groups. Draw a model that represents the equal groups.

How many whole blocks are in each group? _____

How many tenths are in each group? _____

So, 3.6 ÷ 3 = _____.

Check Use multiplication to check your answer.

$$\begin{array}{r} \square . \square \\ \times \quad 3 \\ \hline \square . \square \end{array}$$

Try It

Find 4.8 ÷ 2 using models.

1 The model shows 4.8 using four wholes and eight tenths.

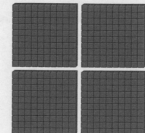 

2 Divide the blocks into two equal groups. Draw a model to represent the equal groups.

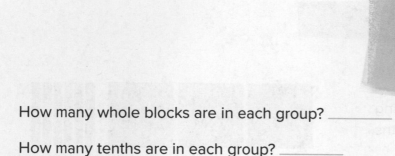

How many whole blocks are in each group? _____

How many tenths are in each group? _____

So, 4.8 ÷ 2 = _____ .

Check Use multiplication to check your answer.

$$\begin{array}{r} \boxed{}.\boxed{} \\ \times \qquad 2 \\ \hline \boxed{}.\boxed{} \end{array}$$

Talk About It

1. The table shows some quotients. Study the table. Write a rule you can use to find the quotient of a decimal and a whole number without using models.

Decimals	Whole Numbers
3.6 ÷ 3 = 1.2	36 ÷ 3 = 12
4.8 ÷ 2 = 2.4	48 ÷ 2 = 24
2.4 ÷ 4 = 0.6	24 ÷ 4 = 6

_____

2. **Processes & Practices** **3** **Justify Conclusions** Use your rule from Exercise 1 to find 3.5 ÷ 7 without using models. Explain the process you used.

_____

_____

Practice It

Use models to find each quotient. Draw the equal groups.

3. $3.4 \div 2 =$ _____

4. $6.3 \div 3 =$ _____

5. $5.6 \div 4 =$ _____

6. $2.7 \div 3 =$ _____

7. Marketa used 4.5 cups of sugar for 5 batches of cookies. If an equal amount was used for each batch, how many cups of sugar were used for each batch? Draw models to help you divide.

_____

8. Alexander ran 3.2 miles over the past 2 days. If he ran an equal amount each day, how many miles did he run each day? Draw models to help you divide.

_____

Processes &Practices [3]

9. **Find the Error** Joseph used base-ten blocks to find 2.1 ÷ 3. He stated that 2.1 ÷ 3 = 0.6. Find his mistake and correct it.

_____

_____

_____

Write About It

10. How can I use models to divide decimals by whole numbers?

_____

_____

_____

Name _____

 MY Homework

Homework Helper

Need help? connectED.mcgraw-hill.com

Find 2.8 ÷ 4 using models.

1 The model shows 2.8 using two wholes and eight tenths.

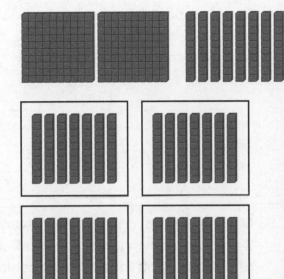

2 The model was divided into four equal groups.

There are no whole blocks in each group.

There are 7 tenths in each group.

So, 2.8 ÷ 4 = 0.7.

Check Use multiplication to check your answer.

$$
\begin{array}{r}
\overset{2}{0.7} \\
\times\ 4 \\
\hline
2.8
\end{array}
$$

Practice

Use models to find each quotient. Draw the equal groups.

1. 1.6 ÷ 2 = _____

2. 2.4 ÷ 4 = _____

3. Elaine and her 3 friends purchased snacks after school for $3.60. If each person paid an equal amount, how much did each person pay? Draw models to help you divide.

4. Vaughn spent 3.5 hours at the pool over the past week. If he spent an equal amount of time over the last 7 days, how much time did he spend each day at the pool? Draw models to help you divide.

5. Mora was preparing for 3 achievement assessments. She spent 3.6 hours studying over the weekend. If she spent an equal amount of time studying for each assessment, how much time did she spend on each assessment? Draw models to help you divide.

6. **Processes &Practices** ⑤ **Use Math Tools** Faith helped pull weeds, trim shrubs, and mow the lawn over the past 1.8 hours. If she spent an equal amount of time on each activity, how much time did she spend on each activity? Draw models to help you divide.

Lesson 11
Divide Decimals by Whole Numbers

ESSENTIAL QUESTION
How is multiplying and dividing decimals similar to multiplying and dividing whole numbers?

 Math in My World Watch ▶ Tutor 💬

Example 1

There are 8.4 meters left on a roll of ribbon. Nancy wants to cut the ribbon in half. What will be the size of each piece of ribbon?

Find 8.4 ÷ 2.

Estimate 8 ÷ 2 = _____

1 Place the decimal point directly above the decimal point in the dividend.

$$2\overline{)8\ .\ 4}$$

2 Divide as with whole numbers.

So, 8.4 ÷ 2 = _____ . Each piece of ribbon will be _____ meters long.

Check Use multiplication. 4.2 × 2 = _____

Use models.

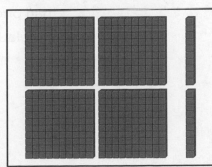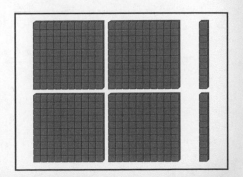

Example 2

Find $6.54 ÷ 12. Round to the nearest cent.

 Place the decimal point directly above the decimal point in the dividend.

 Divide as with whole numbers.

Annex a zero after 6.54 and continue dividing.

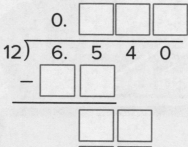

So, $6.54 ÷ 12 = $_____. Rounded to the nearest cent, this is $0.55.

Check Use multiplication.

```
    0.545
  ×    12
    1090
  + 5450
    6.540
```

Guided Practice

Divide.

1. $3\overline{)48.33}$

2. $2\overline{)8.8}$

Talk MATH

Is the quotient of 9.3 ÷ 15 greater than one or less than one? Explain without calculating.

Independent Practice

Divide. Check your answer using multiplication.

3. $145.8 \div 12 =$ _____

4. $22.11 \div 11 =$ _____

5. $38.4 \div 16 =$ _____

6. $8\overline{)12.4}$

7. $14\overline{)14.14}$

8. $11\overline{)55.44}$

Divide. Round to the nearest tenth.

9. $7.21 \div 7 =$ _____

10. $6.28 \div 4 =$ _____

11. $5\overline{)276.2}$

Divide. Round to the nearest hundredth.

12. $78.04 \div 8 =$ _____

13. $24\overline{)75.48}$

14. $25\overline{)4.60}$

Problem Solving

15. Mandy and 5 of her friends bought a package of bottled water for $4.98. How much will each friend pay to the nearest cent, if the cost is divided equally?

_____

16. Four girls swam the 4-by-200 meter freestyle relay in a total of 540.4 seconds. If each girl swam her part of the race in the same amount of time, what was the time for one girl?

_____

17. **Processes &Practices** 2 **Reason** The table shows the prices for two packs of batteries. Which pack is a better buy? Explain your reasoning.

Batteries		
Pack	Number of Batteries	Price
A	4	$2.44
B	6	$3.90

_____

_____

Brain Builders

18. The cost of 4 evening movie tickets is $33.40. The cost of 6 daytime tickets is $39.30. What is the difference between the cost of one evening ticket and one daytime ticket?

_____

19. **Processes &Practices** 4 **Model Math** Write a real-life problem that involves dividing a decimal by a whole number. Include a model to show the solution.

_____

_____

20. ❓ **Building on the Essential Question** When might you need to round the quotient when dividing a decimal by a whole number?

_____

_____

_____

Copyright © McGraw-Hill Education

Name

MY Homework

Homework Helper

Need help? connectED.mcgraw-hill.com

Find 15.3 ÷ 3.

Estimate 15 ÷ 3 = 5

1 Place the decimal point directly above the decimal point in the dividend.

2 Divide as with whole numbers.

$$
\begin{array}{r}
5.1 \\
3\overline{)15.3} \\
-15 \\
\hline
0\ 3 \\
-\ 3 \\
\hline
0
\end{array}
$$

So, 15.3 ÷ 3 = 5.1.

Check Use multiplication to check your answer.   5.1 × 3 = 15.3

Practice

Divide. Check your answer using multiplication.

1. 223.6 ÷ 40 = _____

2. 8.14 ÷ 20 = _____

3. 361.5 ÷ 12 = _____

Divide. Check your answer using multiplication.

4. 2)26.48

5. 5)7.75

6. 3)6.9

Problem Solving

7. Four girls run a relay and finish the race in 6.48 minutes. If each girl ran at the same speed, how many minutes did each girl run? Round to the nearest tenth.

_____

Processes
8. &Practices **6** **Explain to a Friend** Is the quotient 3.6 ÷ 9 greater than or less than 1? Explain to a friend.

_____

_____

Brain Builders

9. A paper company wants to ship 110.4 tons of paper in 4 trucks. If each truck can carry 25 tons of paper, do they have enough trucks?

_____

10. Test Practice Marcel had drama practice 8.75 hours from Monday to Friday. If he practices the same amount each day, how many hours does Marcel spend in drama practice each day?

Ⓐ 1.75 hours

Ⓒ 2.25 hours

Ⓑ 2 hours

Ⓓ 2.50 hours

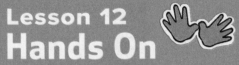

Lesson 12
Hands On
Use Models to Divide Decimals

ESSENTIAL QUESTION
How is multiplying and dividing decimals similar to multiplying and dividing whole numbers?

You can use models to divide decimals by decimals.

How many groups of 8 tenths are in 2.4? Find 2.4 ÷ 0.8.

1 The model shows 2.4.

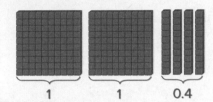

           1   1   0.4

2 Since you are dividing by tenths, replace both of the wholes with tenths.

Each whole equals how many tenths?

_____

How many tenths do you have altogether after replacing the two

wholes with tenths? _____

3 Separate the tenths into groups of eight tenths to show dividing by 0.8.

How many groups will you have?

_____

Are there any left over? _____
Draw the result at the right.

So, 2.4 ÷ 0.8 = _____.

Check Use multiplication to check your answer.

     0.8 × 3 = 2.4

Try It

How many groups of 6 hundredths are in 3 tenths? Find 0.3 ÷ 0.06.

 The decimal model shows 0.3.

How many hundredths are in 3 tenths? _____

 Separate these hundredths into groups of 6 hundredths. Draw the equal groups.

How many groups will you have? _____

Are there any left over? _____

So, 0.3 ÷ 0.06 = _____.

Talk About It

Processes & Practices

1. **8** **Look for a Pattern** The table shows some quotients. Study the table. Write a rule you can use to find the quotient of two decimals, both to the tenths place, without using models.

Decimals	Whole Numbers
2.4 ÷ 0.8 = 3	24 ÷ 8 = 3
3.6 ÷ 0.9 = 4	36 ÷ 9 = 4
2.8 ÷ 0.4 = 7	28 ÷ 4 = 7

_____

_____

Processes & Practices

2. **3** **Draw a Conclusion** Use your rule from Exercise 1 to find 4.8 ÷ 0.8 without using models. Explain the process you used.

_____

_____

Practice It

Use models to find each quotient. Draw the equal groups.

3. $2.4 \div 0.3 =$ _____

4. $1.6 \div 0.4 =$ _____

5. $0.1 \div 0.05 =$ _____

6. $0.2 \div 0.04 =$ _____

7. $0.27 \div 0.03 =$ _____

Processes & Practices 5

8. **Use Math Tools** Grace purchased some packages of gum for $1.50. If each package costs $0.50, how many packages did Grace purchase? Use models to help you divide.

_____

9. Aidan purchased 1.5 feet of wood to make a napkin holder. If he used 0.5 foot of wood for each side, how many sides does the napkin holder have? Use models to help you divide.

_____

10. Ryan used 2.5 cups of flour to make bread. If he used 0.5 cup of flour for each loaf, how many loaves of bread did Ryan make? Use models to help you divide.

_____

Processes & Practices 6

11. **Explain to a Friend** Explain how to use models to divide 2.1 by 0.7.

_____

_____

_____

Write About It

12. How can I use models to divide decimals by decimals?

_____

_____

_____

MY Homework

Homework Helper

Need help? connectED.mcgraw-hill.com

How many groups of 7 tenths are in 1.4? Find 1.4 ÷ 0.7.

1 The model shows 1.4 using one whole and four tenths.

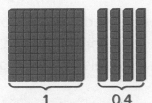

2 Since you are dividing by tenths, the whole block was replaced with tenths.

There are a total of 14 tenths altogether after replacing the one whole with ten tenths.

3 The tenths are separated into groups of seven tenths to show dividing by 0.7.

There are 2 groups with none left over.

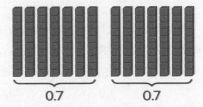

So, 1.4 ÷ 0.7 = 2.

Practice

Use models to find each quotient. Draw the equal groups.

1. 0.4 ÷ 0.05 = _____

2. 0.2 ÷ 0.02 = _____

![Real World] **Problem Solving**

Processes
3. **&Practices** [5] **Use Math Tools** Adam used 1.8 pieces of paper to make his project. He decorated 0.2 of each piece of paper with a different color. How many sections does Adam's art project have? Draw models to help you divide.

_____

4. Beverly used 2.4 yards of ribbon on the border of a quilt. If each side used 0.6 yard of ribbon, how many sides did she use the ribbon on? Draw models to help you divide.

_____

5. Melanie purchased some packages of pencils for $1.60. If each package costs $0.80, how many packages did Melanie purchase? Draw models to help you divide.

_____

6. Evan had 2.2 pounds of candy to give to his friends. If he gave 0.2 pound to each friend, how many friends did Evan give candy to? Draw models to help you divide.

_____

Copyright © McGraw-Hill Education

454 **Need more practice?** Download Extra Practice at **connectED.mcgraw-hill.com**

Name _____

Lesson 13
Divide Decimals

When dividing by decimals, change the divisor into a whole number. To do this, multiply both the divisor and the dividend by the same power of 10. Then divide as with whole numbers.

 Math in My World Watch ▶ Tutor 💬

Example 1

Kasey is 4.5 feet tall. Her brother, Jerome, is 6.75 feet tall. How many times taller is Jerome than Kasey?

Find 6.75 ÷ 4.5.

1 Multiply 4.5 by 10 to make 45. Then, multiply 6.75 by the same number, 10, to make 67.5.

$$4.5\overline{)6.75}$$

$$
\begin{array}{r}
1.5 \\
45\overline{)67.5} \\
-45 \\
\hline
225 \\
-225 \\
\hline
0
\end{array}
$$

2 Place the decimal point in the quotient. Divide as with whole numbers.

So, Jerome is _____ times taller than Kasey.

Check Multiply to check your answer.

$$
\begin{array}{r}
1.\;5 \\
\times\,4.\;5 \\
\hline
\square\,.\square\,\square
\end{array}
$$

Example 2

Find 0.06 ÷ 1.5.

1 Multiply each number by 10.

$$1.5\overline{)0.06}$$

2 Place the decimal point in the quotient. Divide as with whole numbers. Annex zeros in the dividend as needed.

15 does not go into 6, so write a 0 in the hundredths place.

$$
\begin{array}{r}
\square.\square\square \\
15\overline{)0.6\ \ 0} \\
0 \\
\hline
6\ \ 0 \\
-\ \square\square \\
\hline
\square
\end{array}
$$

Check Multiply to check your answer.

$$
\begin{array}{r}
0.04 \\
\times\ \ 1.5 \\
\hline
0.06
\end{array}
$$

Guided Practice

1. Divide. Check your answer using multiplication.

$$6.89 \div 1.3 = \underline{\hphantom{XXX}}$$

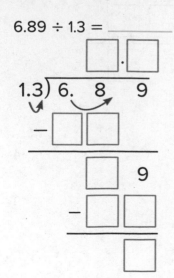

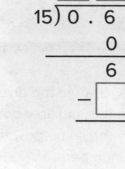

Talk MATH

When finding 0.808 ÷ 0.4, by what number should you multiply the divisor? Explain.

Independent Practice

Divide. Check your answer using multiplication.

2. $0.66 \div 0.3 =$ _____

3. $16.5 \div 0.03 =$ _____

4. $0.462 \div 0.2 =$ _____

5. $0.12\overline{)18.6}$

6. $1.4\overline{)0.07}$

7. $0.04\overline{)3.822}$

8. $14.4 \div 0.4 =$ _____

9. $0.78\overline{)3.51}$

10. $0.15\overline{)84.78}$

Algebra Find each unknown.

11. $0.08 \div 1.6 = z$

$z =$ _____

12. $13.2 \div 0.3 = d$

$d =$ _____

13. $0.92 \div 0.4 = q$

$q =$ _____

Problem Solving

14. **Processes &Practices** **4** **Model Math** Mrs. Chibas is making chocolate chip cookies for her daughter's class. She bought a tub of chocolate chip cookie dough that contained 57.6 ounces of dough. If each cookie needs 1.2 ounces of dough, how many cookies can she make?

_____

15. Elliot ran 4 laps in 209.2 seconds. If he ran at the same speed for each lap, what was his time for each lap?

_____

Brain Builders

16. Kaya works for a T-shirt company and has 49.5 yards of fabric to make specialty T-shirts. Each T-shirt needs 4.5 yards of fabric and sells for $22.50. How much money can Kaya make using the fabric she has?

_____

17. **Processes &Practices** **6** **Be Precise** Without solving, would $1.98 \div 0.51$ be closer to 4 or 5? Explain. Draw a model to support your claim?

_____

_____

18. **?** **Building on the Essential Question** How is dividing decimals different from dividing whole numbers?

_____

_____

_____

_____

Name ....

MY Homework

Homework Helper

Need help? connectED.mcgraw-hill.com

The Prease family drove 213.9 miles to the beach. Their car used 9.3 gallons of gas. How many miles did they drive per gallon of gas?

Find 213.9 ÷ 9.3.

 Multiply 9.3 by 10 to make 93. Then, multiply 213.9 by the same number, 10, to make 2,139.

$$9.3\overline{)213.9}$$

$$\begin{array}{r} 23. \\ 93\overline{)2,139.} \\ -186 \\ \hline 279 \\ -279 \\ \hline 0 \end{array}$$

2 Place the decimal point in the quotient. Divide as with whole numbers.

So, they drove 23 miles per gallon of gas.

Check Multiply to check your answer.

$$\begin{array}{r} 23 \\ \times\ 9.3 \\ \hline 213.9 \end{array}$$

Practice

Divide. Check your answer using multiplication.

1. 12.42 ÷ 4.6 = _____   **2.** 0.4)0.242   **3.** 1.404 ÷ 0.45 = _____

4. Lamar spent $16.25 on peanuts. If he bought 2.6 pounds of peanuts, how much does one pound of peanuts cost?

5. Sei paid $10.48 for a number of washcloths. If each washcloth is $1.31, how many washcloths did Sei get?

Brain Builders

6. **Processes &Practices** 6 **Be Precise** Write and solve a real-world problem that involves dividing a decimal by a decimal. Show the solution and check your answer.

7. Patricia is making two sizes of hamburgers: 0.5-pound and 0.25-pound. How many of each size of hamburger could she make with 3.75 pounds of hamburger?

8. **Test Practice** Mason has 1.71 meters of string to decorate his locker. He cut the string into 0.19-meter lengths. How many pieces does he have?

- Ⓐ 6 pieces
- Ⓑ 7 pieces
- Ⓒ 8 pieces
- Ⓓ 9 pieces

Lesson 14
Divide Decimals by Powers of Ten

ESSENTIAL QUESTION
How is multiplying and dividing decimals similar to multiplying and dividing whole numbers?

You can divide decimals by powers of ten, such as 10, 100, and 1,000. Powers of ten can be written with exponents.

 Math in My World

Example 1

Laura needs to make 10 payments to pay for her new motorized scooter. How much will each payment be if the scooter costs $219.50?

One Way Divide.

Find $219.50 ÷ 10.

```
        21.95
   10 ) 219.50
      − 20
        19
       − 10
          9 5
        − 9 0
            50
          − 50
             0
```

> The non-zero digits are the same. The decimal point in the quotient moved one place to the left.

Another Way Move the decimal point.

Dividing a decimal by 10 moves the decimal point one place to the left.

$219,5 \rightarrow 21.95$

So, each payment will be _____.

Check Use multiplication. 21.95 × 10 = 219.5

To divide a decimal by a power of ten, move the decimal point to the left the same number of zeros as is in the power of ten. This is also the same number as the exponent on 10.

Example 2

In the past 10^2 years, scientists measured the movement of the continents to be 190.5 centimeters. If the continents moved the same amount each year, how much did the continents move in one year?

Helpful Hint

The quotient of a number and a power of ten that is greater than 1 will always be less than the original number.

Since $10^2 = 100$, find $190.5 \div 100$.

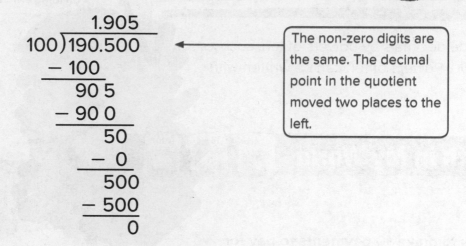

$$\begin{array}{r} 1.905 \\ 100\overline{)190.500} \\ -\ 100 \\ \hline 90\ 5 \\ -\ 90\ 0 \\ \hline 50 \\ -\ 0 \\ \hline 500 \\ -\ 500 \\ \hline 0 \end{array}$$

The non-zero digits are the same. The decimal point in the quotient moved two places to the left.

Dividing a decimal by 100 moved the decimal point two places to the left.

$$190.5 \longrightarrow 1.905$$

So, $190.5 \div 10^2 =$ _____. The continents moved _____ centimeters in one year.

Check Use multiplication. $1.905 \times 100 = 190.5$

Guided Practice

Divide. Check your answer using multiplication.

1. $7.2 \div 10 =$ _____

2. $21.5 \div 10^2 =$ _____

3. $19.2 \div 10^3 =$ _____

Talk MATH

Use the number of zeros in the number 10 to explain why $4.5 \div 10 = 0.45$.

Independent Practice

4. 5.62 ÷ 100 = _____

5. 18.7 ÷ 100 = _____

6. 6.3 ÷ 10^3 = _____

7. 0.05 ÷ 1 = _____

8. 0.012 ÷ 10^2 = _____

9. 2.46 ÷ 10^1 = _____

10. 8.72 ÷ 100 = _____

11. 98.6 ÷ 10^2 = _____

12. 5.71 ÷ 1 = _____

13. 437 ÷ 1,000 = _____

Problem Solving

14. Herbert bought red beans to make chili to sell at a fundraiser. The total price of the red beans was $13.90. How much did it cost for each can?

Items Purchased	
Item	Quantity
Red beans	10
Shredded cheese	6

15. **Processes &Practices** 4 **Model Math** Sarah was trying to decide what kind of cat food to buy at the store. Smiley Cat costs $7.90 for 10 pounds. Purr Plus costs $0.85 per pound. Which brand is the better buy? Explain your reasoning.

Brain Builders

16. In Alaska, scientists are studying the melting of glaciers. They found that in the past 100 years, the glaciers receded about 590.5 feet. If the glaciers receded at the same rate every year, how much did they recede in ten years? How much did they recede in one year?

17. **Processes &Practices** 3 **Which One Doesn't Belong?** Circle the expression that does not belong with the other three. Rewrite the expression so that it does belong. Explain your reasoning.

$$6.7 \div 10 \quad 4 \div 1{,}000 \quad 0.2 \div 1 \quad 52.1 \div 100$$

18. **Building on the Essential Question** Describe the relationship between the number of places a decimal point is moved to the left and the change in the value of the number. Use a model to explain your answer.

Name ....

MY Homework

Homework Helper

Need help? connectED.mcgraw-hill.com

Find 36.4 ÷ 100.

One Way Divide.

$$
\begin{array}{r}
0.364 \\
100\overline{)36.4} \\
-\ 300 \\
\hline
640 \\
-\ 600 \\
\hline
400 \\
-\ 400 \\
\hline
0
\end{array}
$$

> The non-zero digits are the same. The decimal point in the quotient moved two places to the left.

So, 36.4 ÷ 100 = 0.364.

Check Use multiplication. 0.364 × 100 = 36.4

Another Way Move the decimal point.

Dividing a decimal by 100 moves the decimal point two places to the left.

36.4 ⟶ 0.364

Practice

Divide. Check your answer using multiplication.

1. 8.3 ÷ 100 = _____

2. 208 ÷ 10^2 = _____

3. 0.07 ÷ 1 = _____

4. 32.7 ÷ 1,000 = _____

Problem Solving

Use the table to answer Exercises 5–7.

Location	Number of Parking Spots per Row	Combined Width (m)
Grocery Store	10	31.92
Hardware Store	10	31.96
Mall	10	31.30

5. Each row of the grocery store parking lot has 10 parking spots of equal width. What is the width for each spot?

_____

6. How much space is given for each parking spot at the Mall parking lot if each spot has an equal width?

_____

_____

7. Which location gives the greatest space for each parking spot?

_____

Brain Builders

8. **Processes & Practices** 4 **Model Math** Christy purchased 6.75 pounds of red licorice and 2.37 pounds of black licorice. How much licorice does she need to put in each bag if she divides the total amount into 10 equal-sized bags?

_____

9. **Processes & Practices** 6 **Explain to a Friend** Tell what the difference is when you multiply or divide decimals by powers of ten by moving the decimal point.

_____

_____

_____

10. **Test Practice** Tomas buys a trumpet for $108.90. He will make 10 equal payments to pay for the trumpet. What is the amount of each payment that Tomas will make?

   Ⓐ $1.09   Ⓑ $10.89   Ⓒ $19.90   Ⓓ $108.90

Vocabulary Check

Match each word to its definition. Write your answers on the lines provided.

1. **Commutative Property** _____

   A. The product of any factor and 1 is equal to the factor.

2. compatible numbers _____

   B. Finding the approximate value of a number.

3. **Identity Property** _____

   C. The order in which factors are multiplied does not change the product.

4. **Associative Property** _____

   D. Numbers like 10, 100, and 1,000, because they can be written as 10^1, 10^2, and 10^3.

5. powers of 10 _____

   E. Numbers that are easy to multiply mentally.

6. rounding _____

   F. The grouping of the factors does not change the product.

Concept Check

Estimate each product.

7. $4.80 × 5

8. 3.36 × 4

9. 10.8 × 7

Multiply. Check for reasonableness.

10. 6 × 1.8 = _____

11. 4.6
   <u>× 5</u>

12. 7.16
   <u>× 2.1</u>

Multiply. Tell which property you used.

13. 50 × (20 × 1.3) = _____

14. 2 × (50 × 5.6) = _____

Estimate each quotient.

15. 20.6 ÷ 4

16. 24.3 ÷ 8

17. 118.1 ÷ 10

Divide. Check your answer using multiplication.

18. $102.6 ÷ 10^2 =$ _____

19. 1.8 ÷ 0.08 = _____

20. $6.2\overline{)12.71}$

Problem Solving

21. Kasi bought 7 pounds of mozzarella cheese. Each pound costs $4.29. About how much did he spend altogether?

22. Ariel buys 2 fish tanks at the pet store. Each tank holds 9.7 liters of water. How many liters of water will she need to fill both tanks?

Brain Builders

23. It costs $4.25 for one pound of roast beef and $3.75 for one pound of ham. How much will it cost to purchase 2.5 pounds of roast beef and 3.5 pounds of ham? Round to the nearest cent.

24. Describe the pattern below. Write another pattern that has the same fourth value. Describe your pattern.

$$0.41, 0.82, 1.64, 3.28$$

25. The total bill for a dinner party was $134.03. Marcus pays $20.45 of the bill. The other people at the party divide the remaining amount equally and each pay $12.62. How many other people were at the dinner party?

26. Test Practice Sofia paints a rectangular mural on the wall. The area of the mural is 128.52 square feet. The mural is 12.6 feet wide. What is the height of the mural?

ⓐ 115.92 ft

ⓒ 10.2 ft

ⓑ 105.2 ft

ⓓ 8.6 ft

Reflect

Use what you learned about multiplying and dividing decimals to complete the graphic organizer.

Whole
Numbers

Decimals

ESSENTIAL QUESTION

How is multiplying and dividing decimals similar to multiplying and dividing whole numbers?

Vocabulary

Now reflect on the ESSENTIAL QUESTION **Write your answer below.**

_____

_____

_____

Performance Task

Brain Builders

Healthy Snack

Javier wants to make a snack with the following healthy ingredients. The serving size, total carbohydrates, and dietary fiber are listed for a serving.

	Vanilla Yogurt	Frozen Strawberries	Frozen Raspberries
Serving Size	243.58 grams	151.4 grams	123.56 grams
Carbohydrates	33.56 grams	12.2 grams	15.54 grams
Dietary Fiber	0.002 grams	3.124 grams	7.82 grams

Show all your work to receive full credit.

Part A

Javier will include 1.5 servings of vanilla yogurt, 0.75 servings of frozen strawberries, and 0.75 servings of frozen raspberries. Find the total weight of the serving size of his snack. Show the expression you used to determine the answer.

Part B

Find the total carbohydrates that are in Javier's snack. Show the expression you used to determine the answer.

Part C

Find the total dietary fiber that is in his snack. Show the expression you used to determine the answer.

Part D

If Javier has 1.5 servings of vanilla yogurt, 0.75 servings of frozen strawberries, and 0.75 servings of frozen raspberries, how many total servings does he have in his snack?

Part E

Javier wants to calculate a single serving from the total serving size, carbohydrates, and dietary fiber of his snack. Use your answers from **Parts A, B,** and **C,** together with the total number of servings from **Part D** to fill in the following table.

Serving Size	
Carbohydrates	
Dietary Fiber	

Glossary/Glosario

← Go online for the eGlossary.

Go to the eGlossary to find out more about these words in the following 13 languages:

Arabic • Bengali • Brazilian Portuguese • Cantonese • English • Haitian Creole Hmong • Korean • Russian • Spanish • Tagalog • Urdu • Vietnamese

English	Spanish/Español

acute angle An angle with a measure between 0° and 90°.

ángulo agudo Ángulo que mide entre 0° y 90°.

acute triangle A triangle with three acute angles.

triángulo acutángulo Triángulo con tres ángulos agudos.

algebra A branch of mathematics that uses symbols, usually letters, to explore relationships between quantities.

álgebra Rama de las matemáticas que usa símbolos, generalmente letras, para explorar relaciones entre cantidades.

angle Two rays with a common endpoint.

ángulo Dos semirrectas con un extremo común.

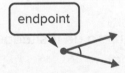

annex To place a zero to the right of a decimal without changing a number's value.

agregar Poner un cero a la derecha de un decimal sin cambiar el valor de un número.

Aa

area   The number of square units needed to cover the surface of a closed figure.

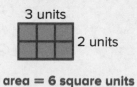

3 units

2 units

area = 6 square units

área   Cantidad de unidades cuadradas necesarias para cubrir la superficie de una figura cerrada.

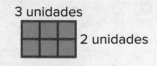

3 unidades

2 unidades

área = 6 unidades cuadradas

Associative Property   Property that states that the way in which numbers are grouped does not change the sum or product.

propiedad asociativa   Propiedad que establece que la manera en que se agrupan los números no altera la suma o el producto.

attribute   A characteristic of a figure.

atributo   Característica de una figura.

axis   A horizontal or vertical number line on a graph. Plural is axes.

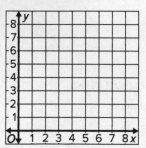

eje   Recta numérica horizontal o vertical en una gráfica.

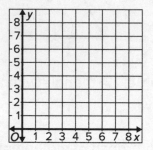

Bb

base   In a power, the number used as a factor. In 10^3, the base is 10.

base   En una potencia, el número que se usa como factor. En 10^3, la base es 10.

base   Any side of a parallelogram.

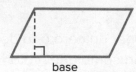

base

base   Cualquiera de los lados paralelogramo.

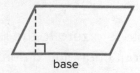

base

base   One of the two parallel congruent faces in a prism.

base   Una de las dos caras congruentes paralelas en un prisma.

capacity   The amount a container can hold.

centimeter (cm)   A metric unit for measuring length.

100 centimeters = 1 meter

common denominator   A number that is a multiple of the denominators of two or more fractions.

common factor   A number that is a factor of two or more numbers.

3 is a common factor of 6 and 12.

common multiple   A whole number that is a multiple of two or more numbers.

24 is a common multiple of 6 and 4.

Commutative Property   Property that states that the order in which numbers are added does not change the sum and that the order in which factors are multiplied does not change the product.

compatible numbers   Numbers in a problem that are easy to work with mentally.

720 and 90 are compatible numbers for division because 72 ÷ 9 = 8.

composite figures   A figure made up of two or more three-dimensional figures.

---

capacidad   Cantidad que puede contener un recipiente.

centímetro (cm)   Unidad métrica de longitud.

100 centímetros = 1 metro

denominador común   Número que es múltiplo de los denominadores de dos o más fracciones.

factor común   Número que es un factor de dos o más números.

3 es factor común de 6 y 12.

múltiplo común   Número natural múltiplo de dos o más números.

24 es un múltiplo común de 6 y 4.

propiedad conmutativa   Propiedad que establece que el orden en que se suman los números no altera la suma y que el orden en que se multiplican los factores no altera el producto.

números compatibles   Números en un problema con los cuales es fácil trabajar mentalmente.

720 ÷ 90 es una divísion que usa son números compatibles porque 72 ÷ 9 = 8.

figura compuesta   Figura conformada por dos o más figuras tridimensionales.

Cc

composite number A whole number that has more than two factors.

12 has the factors 1, 2, 3, 4, 6, and 12.

congruent Having the same measure.

congruent angles Angles of a figure that are equal in measure.

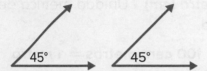

congruent figures Two figures having the same size and the same shape.

congruent sides Sides of a figure that are equal in length.

convert To change one unit to another.

coordinate One of two numbers in an ordered pair.

The 1 is the number on the *x*-axis, the 5 is on the *y*-axis.

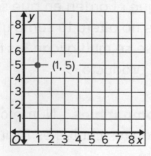

coordinate plane A plane that is formed when two number lines intersect.

número compuesto Número natural que tiene más de dos factores.

12 tiene a los factores 1, 2, 3, 4, 6 y 12.

congruentes Que tienen la misma medida.

ángulos congruentes Ángulos de una figura que tienen la misma medida.

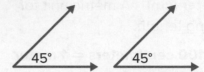

figuras congruentes Dos figuras que tienen el mismo tamaño y la misma forma.

lados congruentes Lados de una figura que tienen la misma longitud.

convertir Transformar una unidad en otra.

coordenada Cada uno de los números de un par ordenado.

El 1 es la coordenada *x* y el 5 es la coordenada *y*.

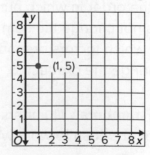

plano de coordenadas Plano que se forma cuando dos rectas numéricas se intersecan formando un ángulo recto.

cube   A rectangular prism with six faces that are congruent squares.

cubo   Prisma rectangular con seis caras que son cuadrados congruentes.

cubed   A number raised to the third power; 10 × 10 × 10, or 10³.

al cubo   Número elevado a la tercera potencia; 10 × 10 × 10 o 10³.

cubic unit   A unit for measuring volume, such as a cubic inch or a cubic centimeter.

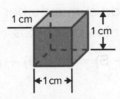

unidad cúbica   Unidad de volumen, como una pulgada cúbica o un centímetro cúbico.

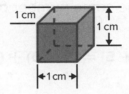

cup   A customary unit of capacity equal to 8 fluid ounces.

taza   Unidad usual de capacidad que equivale a 8 onzas líquidas.

customary system   The units of measurement most often used in the United States. These include foot, pound, and quart.

sistema usual   Conjunto de unidades de medida de uso más frecuente en Estados Unidos. Incluyen el pie, la libra y el cuarto.

Dd

decimal   A number that has a digit in the tenths place, hundredths place, and beyond.

decimal   Número que tiene al menos un dígito en el lugar de las décimas, centésimas etcétera.

decimal point   A period separating the ones and the tenths in a decimal number.

0.8 or $3.77

punto decimal   Punto que separa las unidades y las décimas en un número decimal.

0.8 o $3.77

degree (°)   a. A unit of measure used to describe temperature. b. A unit for measuring angles.

grado (°)   a. Unidad de medida que se usa para describir la temperatura. b. Unidad que se usa para medir ángulos.

Dd

denominator   The bottom number in a fraction. It represents the number of parts in the whole.

In $\frac{5}{6}$, 6 is the denominator.

denominador   Numero que se escribe debajo de la barra en una fracción. Representa el número de partes en que se divide un entero.

En $\frac{5}{6}$, 6 es el denominador.

digit   A symbol used to write numbers. The ten digits are 0, 1, 2, 3, 4, 5, 6, 7, 8, and 9.

dígito   Símbolo que se usa para escribir los números. Los diez dígitos son 0, 1, 2, 3, 4, 5, 6, 7, 8 y 9.

Distributive Property   To multiply a sum by a number, you can multiply each addend by the same number and add the products.

$$8 \times (9 + 5) = (8 \times 9) + (8 \times 5)$$

propiedad distributiva   Para multiplicar una suma por un número, puedes multiplicar cada sumando por ese número y luego sumar los productos.

$$8 \times (9 + 5) = (8 \times 9) + (8 \times 5)$$

divide (division)   An operation on two numbers in which the first number is split into the same number of equal groups as the second number.

12 ÷ 3 means 12 is divided into 3 equal-size groups

dividir (división)   Operación entre dos números en la cual el primer número se separa en tantos grupos iguales como indica el segundo número.

12 ÷ 3 significa que 12 se divide entre 3 grupos de igual tamaño.

dividend   A number that is being divided.

$3\overline{)19}$ ← 19 is the dividend.

dividendo   Número que se divide.

$3\overline{)19}$ ← 19 es el dividendo.

divisible   Describes a number that can be divided into equal parts and has no remainder.

39 is divisible by 3 with no remainder.

divisible   Describe un número que puede dividirse en partes iguales, sin residuo.

39 es divisible entre 3 sin residuo.

divisor   The number that divides the dividend.

3 is the divisor. → $3\overline{)19}$

divisor   Número entre el cual se divide el dividendo.

3 es el divisor. → $3\overline{)19}$

edge The line segment where two faces of a three-dimensional figure meet.

arista Segmento de recta donde se unen dos caras de una figura tridimensional.

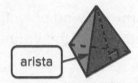

equation A number sentence that contains an equal sign, showing that two expressions are equal.

ecuación Expresión numérica que contiene un signo igual y que muestra que dos expresiones son iguales.

equilateral triangle A *triangle* with three *congruent* sides.

triángulo equilátero *Triángulo* con tres lados *congruentes*.

equivalent decimals Decimals that have the same value.

0.3 and 0.30

decimales equivalentes Decimales que tienen el mismo valor.

0.3 y 0.30

equivalent fractions Fractions that have the same value.

$$\frac{3}{4} = \frac{6}{8} = \frac{9}{12}$$

$\frac{1}{4}$	$\frac{1}{4}$	$\frac{1}{4}$	
$\frac{1}{8}$ $\frac{1}{8}$ $\frac{1}{8}$ $\frac{1}{8}$ $\frac{1}{8}$ $\frac{1}{8}$			
$\frac{1}{12}$ $\frac{1}{12}$ $\frac{1}{12}$ $\frac{1}{12}$ $\frac{1}{12}$ $\frac{1}{12}$ $\frac{1}{12}$ $\frac{1}{12}$ $\frac{1}{12}$			

fracciones equivalentes Fracciones que tienen el mismo valor.

$$\frac{3}{4} = \frac{6}{8} = \frac{9}{12}$$

$\frac{1}{4}$	$\frac{1}{4}$	$\frac{1}{4}$	
$\frac{1}{8}$ $\frac{1}{8}$ $\frac{1}{8}$ $\frac{1}{8}$ $\frac{1}{8}$ $\frac{1}{8}$			
$\frac{1}{12}$ $\frac{1}{12}$ $\frac{1}{12}$ $\frac{1}{12}$ $\frac{1}{12}$ $\frac{1}{12}$ $\frac{1}{12}$ $\frac{1}{12}$ $\frac{1}{12}$			

estimate A number close to an exact value. An estimate indicates about how much.

47 + 22 (round to 50 + 20)

The estimate is 70.

estimación Número cercano a un valor exacto. Una estimación indica una cantidad aproximada.

47 + 22 (se redondea a 50 + 20)

La estimación es 70.

Ee

evaluate   To find the value of an expression by replacing variables with numbers.

even number   A whole number that is divisible by 2.

expanded form   A way of writing a number as the sum of the values of its digits.

exponent   In a power, the number of times the base is used as a factor. In 5^3, the exponent is 3.

expression   A combination of numbers, variables, and at least one operation.

evaluar   Calcular el valor de una expresión reemplazando las variables por números.

número par   Número natural divisible entre 2.

forma desarrollada   Manera de escribir un número como la suma de los valores de sus dígitos.

exponente   En una potencia, el número de veces que se usa la base como factor. En 5^3, el exponente es 3.

expresión   Combinación de números, variables y por lo menos una operación.

Ff

face   A flat surface.

A square is a face of a cube.

cara   Superficia plana.

Cada cara de un cubo es un cuadrado.

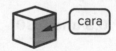

fact family   A group of related facts using the same numbers.

factor   A number that is multiplied by another number.

Fahrenheit (°F)   A unit used to measure temperature.

fair share   An amount divided equally.

fluid ounce   A customary unit of capacity.

familia de operaciones   Grupo de operaciones relacionadas que usan los mismos números.

factor   Número que se multiplica por otro número.

Fahrenheit (°F)   Unidad que se usa para medir la temperatura.

partes iguales   Partes entre las que se divide equitativamente un entero.

onza líquida   Unidad usual de capacidad.

foot (ft)   A customary unit for measuring length. Plural is feet.

1 foot = 12 inches

fraction   A number that represents part of a whole or part of a set.

$$\frac{1}{2}, \frac{1}{3}, \frac{1}{4}, \frac{3}{4}$$

pie (pie)   Unidad usual de longitud.

1 pie = 12 pulgadas

fracción   Número que representa una parte de un todo o una parte de un conjunto.

$$\frac{1}{2}, \frac{1}{3}, \frac{1}{4}, \frac{3}{4}$$

gallon (gal)   A customary unit for measuring capacity for liquids.

1 gallon = 4 quarts

galón (gal)   Unidad de medida usual de capacidad de líquidos.

1 galón = 4 cuartos

gram (g)   A metric unit for measuring mass.

graph   To place a point named by an ordered pair on a coordinate plane.

Greatest Common Factor (GCF)   The greatest of the common factors of two or more numbers.

The greatest common factor of 12, 18, and 30 is 6.

gramo (g)   Unidad métrica para medir la masa.

graficar   Colocar un punto nombrado por un par ordenado en un plano de coordenadas.

máximo común divisor (M.C.D.)   El mayor de los factores comunes de dos o más números.

El máximo común divisor de 12, 18 y 30 es 6.

height   The shortest distance from the base of a parallelogram to its opposite side.

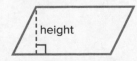

altura   La distancia más corta desde la base de un paralelogramo hasta su lado opuesto.

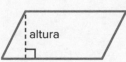

Hh

hexagon   A polygon with six sides and six angles.

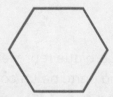

horizontal axis   The axis in a coordinate plane that runs left and right (↔). Also known as the *x*-axis.

hundredth   A place value position. One of one hundred equal parts. In the number 0.57, 7 is in the hundredths place.

hexágono   Polígono con seis lados y seis ángulos.

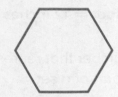

eje horizontal   Eje en un plano de coordenadas que va de izquierda a derecha (↔). También conocido como eje *x*.

centésima   Valor posícional. Una de cien partes iguales. En el número 0.57, 7 está en el lugar de las centésimas.

Ii

Identity Property   Property that states that the sum of any number and 0 equals the number and that the product of any number and 1 equals the number.

improper fraction   A fraction with a numerator that is greater than or equal to the denominator.

$$\frac{17}{3} \text{ or } \frac{5}{5}$$

inch (in.)   A customary unit for measuring length. The plural is inches.

inequality   Two quantities that are not equal.

propiedad de identidad   Propiedad que establece que la suma de cualquier número y 0 es igual al número y que el producto de cualquier número y 1 es igual al número.

fracción impropia   Fracción con un numerador mayor que él igual al denominador.

$$\frac{17}{3} \text{ o } \frac{5}{5}$$

pulgada (pulg)   Unidad usual de longitud.

desigualdad   Dos cantidades que no son iguales.

intersecting lines   *Lines* that meet or cross at a common *point*.

rectas secantes   *Rectas* que se intersecan o se cruzan en un *punto* común.

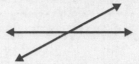

interval   The distance between successive values on a scale.

intervalo   Distancia entre valores sucesivos en una escala.

inverse operations   Operations that undo each other.

operaciones inversas   Operaciones que se cancelan entre sí.

isosceles triangle   A *triangle* with at least 2 *sides* of the same *length*.

triángulo isósceles   *Triángulo* que tiene por lo menos 2 *lados* del mismo largo.

kilogram (kg)   A metric unit for measuring mass.

kilogramo (kg)   Unidad métrica de masa.

kilometer (km)   A metric unit for measuring length.

kilómetro (km)   Unidad métrica de longitud.

Least Common Denominator (LCD) The least common multiple of the denominators of two or more fractions.

$$\frac{1}{12}, \frac{1}{6}, \frac{1}{8}; \text{ LCD is 24.}$$

mínimo común denominador (m.c.d.) El mínimo común múltiplo de los denominadores de dos o más fracciones.

$$\frac{1}{12}, \frac{1}{6}, \frac{1}{8}; \text{ el m.c.d. es 24.}$$

Least Common Multiple (LCM)   The smallest whole number greater than 0 that is a common multiple of each of two or more numbers.

The LCM of 2 and 3 is 6.

mínimo común múltiplo (m.c.m.)   El menor número natural, mayor que 0, múltiplo común de dos o más números.

El m.c.m. de 2 y 3 es 6.

Ll

length Measurement of the distance between two points.

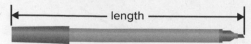

like fractions Fractions that have the same denominator.

$$\frac{1}{5} \text{ and } \frac{2}{5}$$

line A set of *points* that form a straight path that goes on forever in opposite directions.

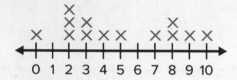

line plot A graph that uses columns of Xs above a number line to show frequency of data.

line segment A part of a *line* that connects two points.

J
K

liter (L) A metric unit for measuring volume or capacity.

1 liter = 1,000 milliliters

longitud Medida de la distancia entre dos puntos.

fracciones semejantes Fracciones que tienen el mismo denominador.

$$\frac{1}{5} \text{ y } \frac{2}{5}$$

recta Conjunto de *puntos* que forman una trayectoria recta sin fin en direcciones opuestas.

diagrama lineal Gráfica que usa columnas de X sobre una recta numérica para mostrar la frecuencia de los datos.

segmento de recta Parte de una *recta* que conecta dos puntos.

J
K

litro (L) Unidad métrica de volumen o capacidad.

1 litro = 1,000 mililitros

Mm

mass Measure of the amount of matter in an object.

meter (m) A metric unit used to measure length.

masa Medida de la cantidad de materia en un cuerpo.

metro (m) Unidad métrica que se usa para medir la longitud.

metric system (SI)   The decimal system of measurement. Includes units such as meter, gram, and liter.

mile (mi)   A customary unit of measure for length.

$$1 \text{ mile} = 5{,}280 \text{ feet}$$

milligram (mg)   A metric unit used to measure mass.

$$1{,}000 \text{ milligrams} = 1 \text{ gram}$$

milliliter (mL)   A metric unit used for measuring capacity.

$$1{,}000 \text{ milliliters} = 1 \text{ liter}$$

millimeter (mm)   A metric unit used for measuring length.

$$1{,}000 \text{ millimeters} = 1 \text{ meter}$$

mixed number   A number that has a whole number part and a fraction part. $3\frac{1}{2}$ is a mixed number.

multiple (multiples)   A multiple of a number is the product of that number and any whole number.

15 is a multiple of 5 because
$3 \times 5 = 15$.

multiplication   An operation on two numbers to find their product. It can be thought of as repeated addition. 4×3 is another way to write the sum of four 3s, which is $3 + 3 + 3 + 3$ or 12.

sistema métrico (SI)   Sistema decimal de medidas que incluye unidades como el metro, el gramo y el litro.

milla (mi)   Unidad usual de longitud.

$$1 \text{ milla} = 5{,}280 \text{ pies}$$

miligramo (mg)   Unidad métrica de masa.

$$1{,}000 \text{ miligramos} = 1 \text{ gramo}$$

mililitro (mL)   Unidad métrica de capacidad.

$$1{,}000 \text{ mililitros} = 1 \text{ litro}$$

milímetro (mm)   Unidad métrica de longitud.

$$1{,}000 \text{ milímetros} = 1 \text{ metro}$$

número mixto   Número formado por un número natural y una parte fraccionaria. $3\frac{1}{2}$ es un número mixto.

múltiplo   Un múltiplo de un número es el producto de ese número por cualquier otro número natural.

15 es múltiplo de 5 porque
$3 \times 5 = 15$.

multiplicación   Operación entre dos números para hallar su producto. También se puede interpretar como una suma repetida. 4×3 es otra forma de escribir la suma de cuatro veces 3, la cual es $3 + 3 + 3 + 3$ o 12.

Nn

net   A two-dimensional pattern of a three-dimensional figure.

number line   A line that represents numbers as points.

numerator   The top number in a fraction; the part of the fraction that tells the number of parts you have.

numerical expression   A combination of numbers and operations.

modelo plano   Patrón bidimensional de una figura tridimensional.

recta numérica   Recta que representa números como puntos.

numerador   Número que se escribe sobre la barra de fracción; la parte de la fracción que indica el número de partes que hay.

expresión numérica   Combinación de números y operaciones.

Oo

obtuse angle   An angle that measures between 90° and 180°.

obtuse triangle   A *triangle* with one *obtuse angle*.

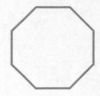

octagon   A polygon with eight sides.

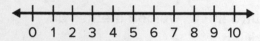

ángulo obtuso   Ángulo que mide entre 90° y 180°.

triángulo obtusángulo   *Triángulo* con un *ángulo obtuso.*

octágono   Polígono de ocho lados.

odd number   A number that is not divisible by 2; such a number has 1, 3, 5, 7, or 9 in the ones place.

número impar   Número que no es divisible entre 2. Los números impares tienen 1, 3, 5, 7, o 9 en el lugar de las unidades.

order of operations   A set of rules to follow when more than one operation is used in an expression.
1. Perform operations in parentheses.
2. Find the value of exponents.
3. Multiply and divide in order from left to right.
4. Add and subtract in order from left to right.

orden de las operaciones   Conjunto de reglas a seguir cuando se usa más de una operación en una expresión.
1. Realiza las operaciones dentro de los paréntesis.
2. Halla el valor de las potencias.
3. Multiplica y divide de izquierda a derecha.
4. Suma y resta de izquierda a derecha.

ordered pair   A pair of numbers that is used to name a point on the coordinate plane.

par ordenado   Par de números que se usa para nombrar un punto en un plano de coordenadas.

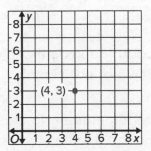

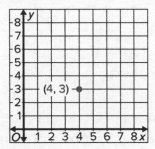

origin   The point (0, 0) on a coordinate plane where the vertical axis meets the horizontal axis.

origen   El punto (0, 0) en un plano de coordenadas donde el eje vertical interseca el eje horizontal.

ounce (oz)   A customary unit for measuring weight or capacity.

onza (oz)   Unidad usual de peso o capacidad.

parallel lines   Lines that are the same distance apart. Parallel lines do not meet.

rectas paralelas   Rectas separadas por la misma distancia en cualquier punto. Las rectas paralelas no se intersecan.

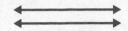

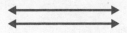

parallelogram   A quadrilateral with four sides in which each pair of opposite sides are parallel and congruent.

partial quotients   A method of dividing where you break the dividend into sections that are easy to divide.

pentagon   A polygon with five sides.

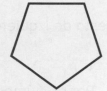

perimeter   The *distance* around a polygon.

period   Each group of three digits on a place-value chart.

perpendicular lines   Lines that meet or cross each other to form right angles.

pint (pt)   A customary unit for measuring capacity.

1 pint = 2 cups

place   The position of a digit in a number.

place value   The value given to a digit by its position in a number.

place-value chart   A chart that shows the value of the digits in a number.

paralelogramo   Cuadrilátero en el cual cada par de lados opuestos son paralelos y congruentes.

cocientes parciales   Método de división por el cual se descompone el dividendo en secciones que son fáciles de dividir.

pentágono   Polígono de cinco lados.

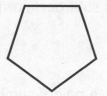

perímetro   *Distancia* alrededor de un polígono.

período   Cada grupo de tres dígitos en una tabla de valor posicional.

rectas perpendiculares   Rectas que se cruzan formando ángulos rectos.

pinta (pt)   Unidad usual de capacidad.

1 pinta = 2 tazas

posición   Lugar que ocupa un dígito en un número.

valor posicional   Valor dado a un dígito según su posición en el número.

tabla de valor posicional   Tabla que muestra el valor de los dígitos en un número.

plane   A flat surface that goes on forever in all directions.

point   An exact location in space that is represented by a dot.

polygon   A closed figure made up of line segments that do not cross each other.

positive number   Number greater than zero.

pound (lb)   A customary unit for measuring weight or mass.

$$1 \text{ pound} = 16 \text{ ounces}$$

power   A number obtained by raising a base to an exponent.

$$5^2 = 25 \quad \textbf{25 is a power of 5.}$$

power of 10   A number like 10, 100, 1,000 and so on. It is the result of using only 10 as a factor.

prime factorization   A way of expressing a composite number as a product of its prime factors.

prime number   A whole number with exactly two factors, 1 and itself.

7, 13, and 19

prism   A three-dimensional figure with two parallel, congruent faces, called bases. At least three faces are rectangles.

product   The answer to a multiplication problem.

plano   Superficie plana que se extiende infinitamente en todas direcciones.

punto   Ubicación exacta en el espacio que se representa con una marca puntual.

polígono   Figura cerrada compuesta por segmentos de recta que no se intersecan.

número positivo   Número mayor que cero.

libra (lb)   Unidad usual de peso.

$$1 \text{ libra} = 16 \text{ onzas}$$

potencia   Número que se obtiene elevando una base a un exponente.

$$5^2 = 25 \quad \textbf{25 es una potencia de 5.}$$

potencia de 10   Número como 10, 100, 1,000, etc. Es el resultado de solo usar 10 como factor.

factorización prima   Manera de escribir un número compuesto como el producto de sus factores primos.

número primo   Número natural que tiene exactamente dos factores: 1 y sí mismo.

7, 13 y 19

prisma   Figura tridimensional con dos caras congruentes y paralelas llamadas bases. Al menos tres caras son rectangulares.

producto   Repuesta a un problema de multiplicación.

Pp

proper fraction A fraction in which the numerator is less than the denominator.

$$\frac{1}{2}$$

property A rule in mathematics that can be applied to all numbers.

protractor A tool used to measure and draw angles.

fracción propia Fracción en la que el numerador es menor que el denominador.

$$\frac{1}{2}$$

propiedad Regla de las matemáticas que puede aplicarse a todos los números.

transportador Instrumento que se usa para medir y trazar ángulos.

Qq

quadrilateral A polygon that has 4 sides and 4 angles.

cuadrilátero Polígono con 4 lados y 4 ángulos.

quart (qt) A customary unit for measuring capacity.

1 quart = 4 cups

cuarto (ct) Unidad usual de capacidad.

1 cuarto = 4 tazas

quotient The result of a division problem.

cociente Resultado de un problema de división.

Rr

ray A line that has one endpoint and goes on forever in only one direction.

semirrecta Parta de una recta que tiene un extremo que se extiende infinitamente en una sola dirección.

rectangle A quadrilateral with four right angles; opposite sides are equal and parallel.

rectángulo Cuadrilátero con cuatro ángulo rectos; los lados opuestos son iguales y paralelos.

rectangular prism A prism that has six rectangular bases.

prisma rectangular Prisma que tiene bases rectangulares.

regular polygon   A polygon in which all sides are congruent and all angles are congruent.

polígono regular   Polígono que tiene todos los lados congruentes y todos los ángulos congruentes.

remainder   The number that is left after one whole number is divided by another.

residuo   Número que queda después de dividir un número natural entre otro.

rhombus   A *parallelogram* with four *congruent sides*.

rombo   *Paralelogramo* con cuatro *lados congruentes*.

right angle   An angle with a measure of 90°.

ángulo recto   Ángulo que mide 90°.

right triangle   A *triangle* with one *right angle*.

triángulo rectángulo   *Triángulo* con un *ángulo recto*.

rounding   To find the approximate value of a number.

redondear   Hallar el valor aproximado de un número.

6.38 rounded to the nearest tenth is 6.4.

6.38 redondeado a la décima más cercana es 6.4.

Ss

scale   A set of numbers that includes the least and greatest values separated by equal intervals.

escala   Conjunto de números que incluye los valores menor y mayor separados por intervalos iguales.

scalene triangle   A *triangle* with no *congruent sides*.

triángulo escaleno   *Triángulo* sin *lados congruentes*.

Ss

scaling The process of resizing a number when it is multiplied by a fraction that is greater than or less than 1.

simplificar Proceso de redimensionar un número cuando se multiplica por una fracción que es mayor que o menor que 1.

sequence A list of numbers that follow a specific pattern.

secuencia Lista de números que sigue un patrón específico.

simplest form A fraction in which the GCF of the numerator and the denominator is 1.

forma simplificada Fracción en la cual el M.C.D. del numerador y del denominador es 1.

solution The value of a variable that makes an equation true. The solution of $12 = x + 7$ is 5.

solución Valor de una variable que hace que la ecualción sea verdadera. La solución de $12 = x + 7$ es 5.

solve To replace a variable with a value that results in a true sentence.

resolver Remplazar una variable por un valor que hace que la expresión sea verdadera.

square A rectangle with four *congruent* sides.

cuadrado Rectángulo con cuatro *lados congruentes.*

square number A number with two identical factors.

número al cuadrado Número con dos factores idénticos.

square unit A unit for measuring area, such as square inch or square centimeter.

unidad cuadrada Unidad de área, como una pulgada cuadrada o un centímetro cuadrado.

squared A number raised to the second power; 3×3, or 3^2.

al cuadrado Número elevado a la segunda potencia; 3×3 o 3^2.

standard form The usual or common way to write a number using digits.

forma estándar Manera usual o común de escribir un número usando dígitos.

straight angle An angle with a measure of 180°.

ángulo llano Ángulo que mide 180°.

sum   The answer to an addition problem.

suma   Respuesta que se obtiene al sumar.

Tt

tenth   A place value in a decimal number or one of ten equal parts or $\frac{1}{10}$.

décima   Valor posicional en un número decimal o una de diez partes iguales o $\frac{1}{10}$.

term   A number in a pattern or sequence.

término   Cada número en un patrón o una secuencia.

thousandth(s)   One of a thousand equal parts or $\frac{1}{1000}$. Also refers to a place value in a decimal number. In the decimal 0.789, the 9 is in the thousandths place.

milésima(s)   Una de mil partes iguales o $\frac{1}{1000}$. También se refiere a un valor posicional en un número decimal. En el decimal 0.789, el 9 está en el lugar de las milésimas.

three-dimensional figure   A figure that has length, width, and height.

figura tridimensional   Figura que tiene largo, ancho y alto.

ton (T)   A customary unit for measuring weight.   1 ton = 2,000 pounds

tonelada (T)   Unidad usual de peso. 1 tonelada = 2,000 libras

trapezoid   A quadrilateral with exactly one pair of parallel sides.

trapecio   Cuadrilátero con exactamente un par de lados paralelos.

triangle   A polygon with three sides and three angles.

triángulo   Polígono con tres lados y tres ángulos.

triangular prism   A prism that has triangular bases.

prisma triangular   Prisma con bases triangulares.

Uu

unit cube   A cube with a side length of one unit.

cubo unitario   Cubo con lados de una unidad de longitud.

unit fraction   A fraction with 1 as its numerator.

fracción unitaria   Fracción que tiene 1 como su numerador.

unknown   A missing value in a number sentence or equation.

incógnita   Valor que falta en una oración numérica o una ecuación.

unlike fractions   Fractions that have different denominators.

fracciones no semejantes   Fracciones que tienen denominadores diferentes.

variable   A letter or symbol used to represent an unknown quantity.

variable   Letra o símbolo que se usa para representar una cantidad desconocida.

vertex   The point where two rays meet in an angle or where three or more faces meet on a three-dimensional figure.

vértice   a. Punto donde se unen los dos lados de un ángulo. b. Punto en una figura tridimensional donde se intersecan 3 o más aristas.

vertical axis   A vertical number line on a graph (↕). Also known as the *y*-axis.

eje vertical   Recta numérica vertical en una gráfica (↕). También conocido como eje *y*.

volume   The amount of space inside a three-dimensional figure.

volumen   Cantidad de espacio que contiene una figura tridimensional.

weight A measurement that tells how heavy an object is.

peso Medida que indica cuán pesado o liviano es un cuerpo.

x-**axis** The horizontal axis (↔) in a coordinate plane.

eje *x* Eje horizontal (↔) en un plano de coordenadas.

x-**coordinate** The first part of an ordered pair that indicates how far to the right of the *y*-axis the corresponding point is.

coordenada *x* Primera parte de un par ordenado; indica a qué distancia a la derecha del eje *y* está el punto correspondiente.

yard (yd) A customary unit of length equal to 3 feet or 36 inches.

yarda (yd) Unidad usual de longitud igual a 3 pies o 36 pulgadas.

y-**axis** The vertical axis (↕) in a coordinate plane.

eje *y* Eje vertical (↕) en un plano de coordenadas.

y-**coordinate** The second part of an ordered pair that indicates how far above the *x*-axis the corresponding point is.

coordenada *y* Segunda parte de un par ordenado; indica a qué distancia por encima del eje *x* está el punto correspondiente.

Work Mat 1: Number Lines

0 1 2 3 4 5 6 7 8 9 10

0 1 2 3 4 5 6 7 8 9 10 11 12 13 14 15 16 17 18 19 20

Work Mat 2: Place-Value Chart (Billions to Ones)

Ones		hundreds	tens	ones

Thousands		hundreds	tens	ones

Millions		hundreds	tens	ones

Billions		hundreds	tens	ones

Work Mat 3: Centimeter Grid

Work Mat 4: Place-Value Chart (Hundreds to Thousandths)

Decimals			Ones		
thousandths	hundredths	tenths	ones	tens	hundreds

Work Mat 5: Tenths and Hundredths Models

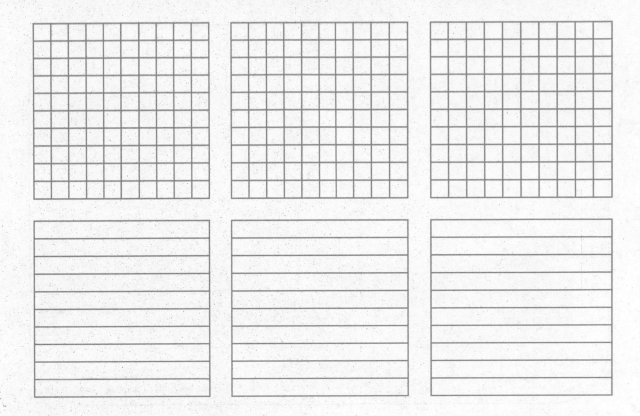

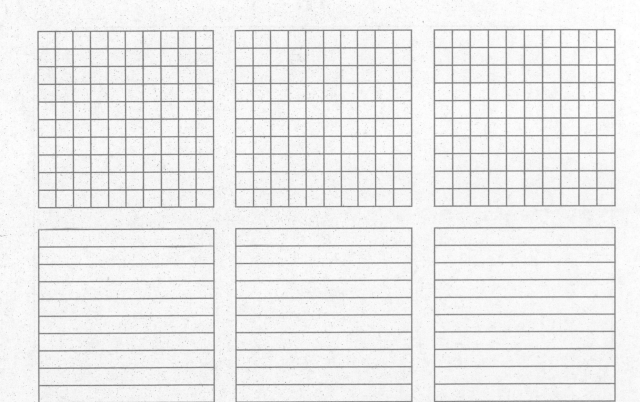

Work Mat 6: Algebra Mat

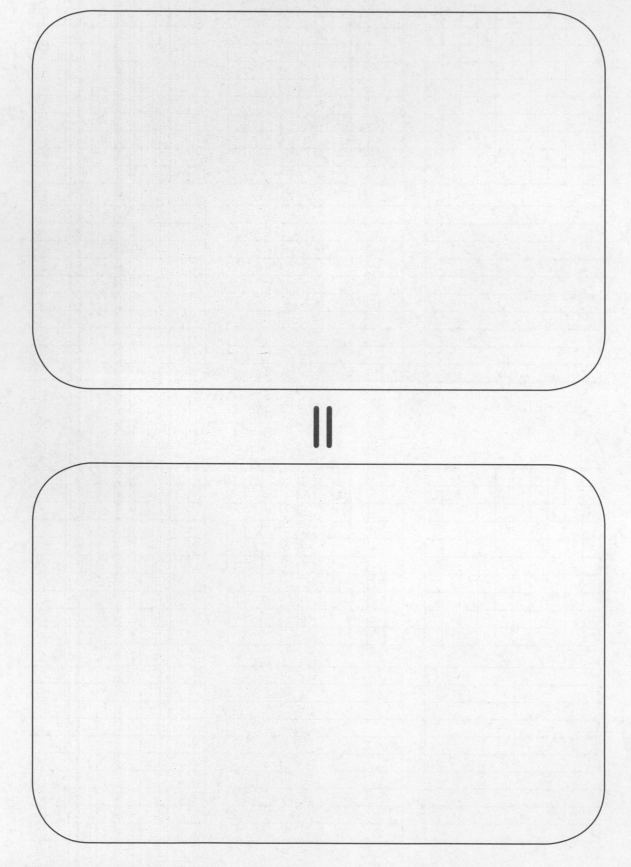

Work Mat 7: First-Quadrant Grid

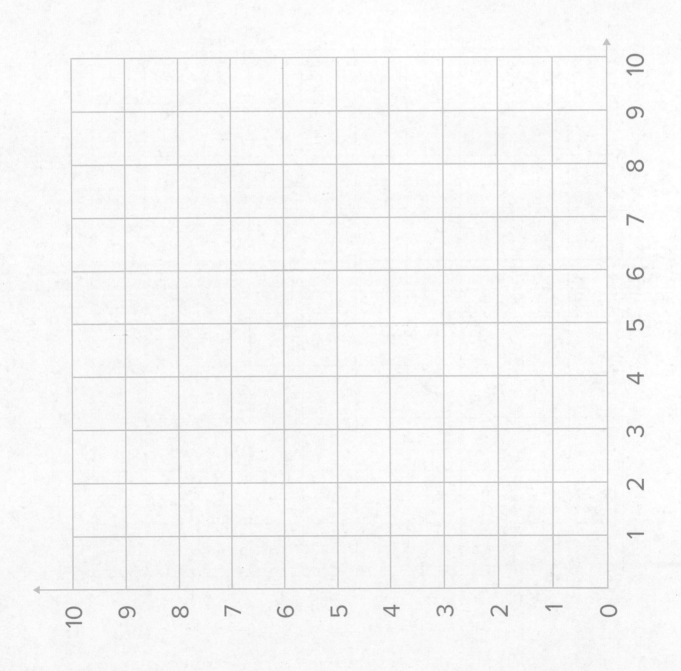

Work Mat 8: First-Quadrant Grid (blank)